Intellectual Property Rights and the Dissemination of Research Tools in Molecular Biology

Summary of a Workshop Held at the
National Academy of Sciences,
February 15

National Research Council

NATIONAL ACADEMY PRESS
Washington, D.C. 1997

NATIONAL ACADEMY PRESS • 2101 Constitution Ave., N.W. • Washington, D.C. 20418

NOTICE: The project that is the subject of this report was approved by the Governing Board of the National Research Council, whose members are drawn from the councils of the National Academy of Sciences, the National Academy of Engineering, and the Institute of Medicine. The members of the committee responsible for the report were chosen for their special competences and with regard for appropriate balance.

This report has been reviewed by a group other than the authors according to procedures approved by a Report Review Committee consisting of members of the National Academy of Sciences, the National Academy of Engineering, and the Institute of Medicine.

The National Academy of Sciences is a private, nonprofit, self-perpetuating society of distinguished scholars engaged in scientific and engineering research, dedicated to the furtherance of science and technology and to their use for the general welfare. Upon the authority of the charter granted to it by the Congress in 1863, the Academy has a mandate that requires it to advise the federal government on scientific and technical matters. Dr. Bruce Alberts is president of the National Academy of Sciences.

The National Academy of Engineering was established in 1964, under the charter of the National Academy of Sciences, as a parallel organization of outstanding engineers. It is autonomous in its administration and in the selection of its members, sharing with the National Academy of Sciences the responsibility for advising the federal government. The National Academy of Engineering also sponsors engineering programs aimed at meeting national needs, encourages education and research, and recognizes the superior achievements of engineers. Dr. William A. Wulf is president of the National Academy of Engineering.

The Institute of Medicine was established in 1970 by the National Academy of Sciences to secure the services of eminent members of appropriate professions in the examination of policy matters pertaining to the health of the public. The Institute acts under the responsibility given to the National Academy of Sciences by its congressional charter to be an adviser to the federal government and, upon its own initiative, to identify issues of medical care, research, and education. Dr. Kenneth I. Shine is president of the Institute of Medicine.

The National Research Council was organized by the National Academy of Sciences in 1916 to associate the broad community of science and technology with the Academy's purposes of furthering knowledge and advising the federal government. Functioning in accordance with general policies determined by the Academy, the Council has become the principal operating agency of both the National Academy of Sciences and the National Academy of Engineering in providing services to the government, the public, and the scientific and engineering communities. The Council is administered jointly by both Academies and the Institute of Medicine. Dr. Bruce Alberts and Dr. Robert M. White are chairman and vice chairman, respectively, of the National Research Council.

This study was sponsored by the National Research Council Basic Science Fund.

International Standard Book Number 0-309-05748-5

Additional copies are available from:

National Academy Press, 2101 Constitution Ave, NW, Box 285, Washington, DC 20055
800-624-6242; 202-334-3313 (in the Washington Metropolitan Area)
http://www.nap.edu

Printed in the United States of America

COMMITTEE ON INTELLECTUAL PROPERTY RIGHTS AND RESEARCH TOOLS IN MOLECULAR BIOLOGY

REBECCA EISENBERG, *(Chair),* University of Michigan Law School, Ann Arbor, Michigan
WILLIAM R. BRINKLEY, Baylor College of Medicine, Houston, Texas
WILLIAM T. COMER, SIBIA, La Jolla, California
BARBARA J. MAZUR, E.I. du Pont de Nemours & Company, Wilmington, Delaware
LITA L. NELSEN, Massachusetts Institute of Technology, Cambridge, Massachusetts
GERALD M. RUBIN, University of California, Berkeley, California
SIDNEY G. WINTER JR., Wharton School of Business, Philadelphia, Pennsylvania

National Research Council Staff

JANET E. JOY, Study Director
ROBIN SCHOEN, Staff Officer
NORMAN GROSSBLATT, Editor
JEFFREY PECK, Project Assistant

VIRTUAL COMMISSION ON INTELLECTUAL PROPERTY RIGHTS

JOHN D. STOBO (*Chair*), Johns Hopkins University School of Medicine, Baltimore, Maryland
SHIRLEY M. TILGHMAN, Princeton University, Princeton, New Jersey
MICHAEL T. CLEGG, University of California, Riverside, California
GERALD D. FISCHBACH, Harvard Medical School, Boston, Massachusetts
WILLIAM HUBBARD JR., Hickory Corners, Michigan
MARY LAKE POLAN, Stanford University School of Medicine, Stanford, California

National Research Council Staff

PAUL GILMAN, Executive Director
SOLVEIG PADILLA, Administrative Assistant

Preface

Although controversy over intellectual property has been a recurring phenomenon in research science, the terms of controversy have shifted in recent years as patenting has become a more familiar part of the landscape. Whereas in an earlier era we might have asked whether intellectual property is fundamentally inconsistent with the norms of research science, today we are likely to ask more nuanced questions about what sorts of research discoveries should be patented and about how proprietary research tools should be disseminated in the research community so as to preserve the benefits of intellectual property while minimizing interference with the progress of science.

In March 1993, in the wake of a controversy over patent applications filed by the National Institutes of Health (NIH) on anonymous cDNA fragments (ESTs), the National Research Council's Commission on Life Sciences (CLS) and the Institute of Medicine (IOM) jointly held a workshop to discuss both specific concerns raised by the NIH patent application and broader issues arising from the patenting of discoveries in the biomedical sciences. Participants at that workshop concluded that a study on intellectual property, technology transfer, and conflict of interest in molecular biology might help resolve some of the issues. In October 1994, NIH director Harold Varmus met with the council of the National Academy of Sciences to discuss how the scientific community should respond to various constraints on the use of research tools and, in particular, to the terms set by Human Genome Sciences for access to its private EST database. In July 1995, the CLS and IOM formed the Committee on Intellectual Property and Research Tools in Molecular Biology to organize a workshop to examine the impact of intellectual property protection on the development of and access to research tools in molecular biology, with attention to the perspectives of universities, government agencies, and private firms.

The workshop was held at the National Academy of Sciences on February 15-16, 1996. Over 100 people attended, coming from academic institutions, industry, and government agencies that sponsor research in molecular biology. The workshop was organized in three sessions. The first consisted of invited papers presenting legal, economic, and sociological perspectives on the topic of intellectual property protection for research tools. The second session consisted of a series of panel discussions of five case studies chosen to illustrate different strategies for managing intellectual property rights in different types of research tools. The third session consisted of presentations of different perspectives from academic institutions (representing both research scientists and technology transfer professionals), industry (representing both small biotechnology companies and major pharmaceutical companies), and government.

This report summarizes the workshop sessions and examines the common themes that emerged. The variety of circumstances presented in the case studies cautions against facile generalizations about ideal practices for protection of research tools in molecular biology. Nonetheless, some themes emerged that might provide useful insights for those concerned with how best to manage intellectual property rights in research tools in molecular biology.

All of the members of the organizing committee gave generously of their time in planning the workshop, identifying speakers, chairing workshop sessions, and reviewing drafts of the report. Francis Collins, director of the National Center for Human Genome Research at NIH and Maxwell Cowan, Vice President and Chief Scientific Officer of the Howard Hughes Medical Institute, met with the committee in advance of the workshop and gave extremely helpful comments and suggestions. The workshop speakers deserve special thanks for the time and thought they put into the workshop. Janet Joy provided essential staff support, both in the planning stages and in drafting the report of the workshop, with the assistance of Jeff Peck. Bob Cook-Deegan provided thoughtful guidance in shaping the final report, and Robin Schoen joined the committee in its planning meeting. Norman Grossblatt edited the proceedings. Last but by no means least, financial support for the workshop came from the National Research Council Basic Science Fund, the Academy Industry Program (AIP) of the National Academy of Sciences, the National Academy of Engineering, and the Institute of Medicine, and the National Institutes of Health.

Rebecca S. Eisenberg, *Chair*

Contents

1

Introduction

Intellectual property rights have been a recurring source of controversy in the biomedical sciences in recent years. A variety of developments have contributed to the increasing salience of intellectual property in biomedical research, including strong and growing commercial interest in the field, legal decisions that have clarified the availability of patent protection for a wide range of discoveries related to life forms, and changes in federal policy to encourage patenting of the results of government-sponsored research.

Protection of intellectual property rights has helped researchers and institutions to attract research funding and has helped firms to raise investment capital and pursue product development. But it has also periodically generated complaints and concerns about its effect on the progress of science and on the dissemination and use of new knowledge. The concerns have been particularly pressing for scientists when intellectual property rights have threatened to restrict access to materials and techniques that are critical for future research. Controversy over intellectual property rights in biomedical research has waxed and waned over the years (Weiner 1989). The current wave of concern was triggered in 1991 when NIH filed its first patent application on partial cDNA sequences, or expressed sequence tags (ESTs). Despite the later withdrawal of the patent applications, the concern over access to DNA sequence information continued to generate debate, both in the US and internationally. Another focal point of concern has been the patenting and licensing of polymerase chain reaction (PCR) technology. In 1992, the pharmaceutical giant Hoffman-La Roche, who holds the patent on the enzyme used in PCR (*Taq* polymerase), sued the biotechnology company Promega for breach of contract over the distribution of enzyme. During the course of the litigation, many research scientists received a letter from Promega suggesting to them they had been named as infringers against the Roche patent. Although Roche stated they had no intention of naming any scientists in the suit, the letter sent a chill throughout the research community and raised fears that patents might be blocking access to research tools. In addition to those controversies, less notorious controversies have surrounded other research tools in molecular biology.

Attitudes toward patenting in the research community have changed substantially since the late 1970s, when Stanford University's decision to patent the recombinant DNA technology developed by Stanley Cohen and Herbert Boyer met considerable resistance among academic scientists. Today, universities and academic scientists routinely pursue patent rights, often in competition with their counterparts in the private sector. University patenting steadily increased from 1965 to about 1980, when there was a sharp increase in patenting that has continued into the 1990s. From 1965 to 1992, university patents increased by a factor of over 15 from 96 to 1500, whereas total patents increased by only about 50% (Henderson and others 1994, 1995). The greatest portion of the increase in university patenting has been in biomedical sciences, and many university patents cover inventions that are useful primarily for scientific research.

There is no clear line separating the interests of the public and private sectors in intellectual property. It is sometimes celebrated and sometimes criticized throughout the research community, not always in the same terms or for the same reasons. University scientists complain that the eagerness of private firms to preserve intellectual property poses a threat to open scientific communication, that the prospect of obtaining patents influences research agendas, that overly broad patents stifle research, and that licensing practices impede access to and use of genetic materials and DNA technology. Yet few scientists today would voice wholesale opposition to patenting itself; scientists' concern is more likely to be how to ensure access to patented inventions on reasonable terms. Representatives of the private sector have a somewhat different list of complaints, including the overeagerness of university technology transfer managers to file patent applications, their overestimation of the value of their intellectual property and the underestimation of the additional investment required to turn a research discovery into a product, and their readiness to grant exclusive, rather than nonexclusive, licenses.

Increasing alliances among academe, industry, and government, driven by a combination of economic and legal changes, have challenged institutions in the public and private sectors to balance their sometimes competing interests in the protection of intellectual property. Over the last two decades, public investment in research has been rewarded by a dazzling series of advances in molecular biology. At the same time, scientists have had to adapt to declines in the growth of public funding to explore these research frontiers. The commercial potential of the advances has motivated the private sector to provide additional resources, and a series of laws, beginning in 1980 with the Bayh-Dole Act, have encouraged the pooling of public and private research funds (see box). This environment has been favorable for the development of small, research-intensive biotechnology companies with close links to universities. Indeed, most of the early biotechnology companies were founded by university professors, and many universities now offer "incubator space" for start-up biotechnology companies working in collaboration with university researchers. Pharmaceutical companies are also increasingly eager to establish collaborations with university researchers.

Changing Direction in Federal Technology Law

In 1980, Congress passed both the Bayh-Dole Act[1] and the Stevenson-Wydler Technology Innovation Act[2]. Together these Acts allowed government contractors, small businesses, and nonprofit organizations to retain certain patent rights in government-sponsored research and permitted the funded entity to transfer the technology to third parties.

The stated intent of Bayh-Dole was to ensure that the patented results of federally-funded research would be broadly and rapidly available for all scientific investigation. Bayh-Dole effectively shifted federal policy from a position of putting the results of government-sponsored research directly into the public domain for use by all, to a pro-patent position that stressed the need for exclusive rights as an incentive for industry to undertake the costly investment necessary to bring new products to market. The policy was based on a belief that private entities, given the incentives of the patent system, would do a better job of commercializing inventions than federal agencies. The Act for the first time established a largely uniform government-wide policy on the treatment of inventions made during federally supported R and D.

Stevenson-Wydler is the basic federal technology law. A principal policy established by that Act is that agencies should ensure the full use of the results of the nation's federal investment in R and D. Another is that the law requires federal laboratories to take an active role in the transfer of federally-owned or originated technology to both state and local governments and to the private sector. Stevenson-Wydler required agencies to establish Offices of Research and Technology Applications at their federal laboratories, and to devote a percentage of their R and D budgets to technology transfer.

References

Tribble, JL. 1995. Gene ownership versus access: meeting the needs. In: Genes for the Future: Discovery, Ownership, Access. *Nat'l Agricultural Biotechnology Council, Ithaca, New York.*

Rudolph, L. 1994. Overview of Federal Technology Transfer. Risk: Health, Safety & Environment *5: 133-142.*

[1]Pub. Law No. 96-517, 6(a), 94 Stat. 3015, 3019-27 (1980).

[2]Pub. Law No. 96-480, 94 Stat. 2311 (1980).

The pervasive intertwining of public and private interests makes molecular biology a particularly useful focal point for considering the effect of intellectual property rights on the dissemination and use of research tools. The potential implications of advances in molecular biology for human health raise the stakes of getting the balance between public and private right, particularly when public attention is riveted on the rising costs of health care. Which approaches to handling intellectual property will best serve the goals of fostering continued scientific discovery and ensuring adequate incentives for commercial development of discoveries to promote human health?

This report seeks to shed light on that question from three angles (following the structure of the workshop itself). The first session presented legal, economic, and sociologic perspectives on the issue of intellectual property protection and access to research tools. Rebecca Eisenberg provides a legal orientation to the problem of intellectual property rights in research tools and reviews legal and commercial developments of the last 15 years that have made intellectual property issues particularly prominent in molecular biology. Drawing from his collaboration with Roberto Mazzoleni, Richard Nelson reviews and critiques various theories about the economic costs and benefits of patenting. Stephen Hilgartner presents his observations on differences in the extent of intellectual property protection in the life sciences and on the role of patents in changing the degree of secrecy among biomedical researchers.

The second section was the heart of the workshop: the case studies. The case studies were chosen to illustrate approaches to the protection of intellectual property that have been used in different contexts by different types of institutions. An important goal of the workshop was to examine the consequences of those approaches from different institutional perspectives. The following cases studies were chosen:

- ***Cohen-Boyer recombinant DNA technology***
  A patented research tool, nonexclusively licensed with low fees.
- ***Partial cDNA sequences, or ESTs***
  Three models for disseminating unpatented research tools.
- ***Polymerase Chain Reaction (PCR) technology and Taq polymerase***
  A patented research tool for which licensing agreements were controversial.
- ***DNA and protein sequencing instrumentations***
  Research tools for which strong patent protection promoted broad access.
- ***Research tools in drug discovery***
  Intellectual property protection of complex biological systems.

By design, the case studies present differences that caution against facile generalizations about ideal practices that should be implemented through uniform rules. Nonetheless, a number of interesting themes emerged, and their emergence suggested that the experience gained through various experiments in technology transfer and the management of intellectual property can, indeed, provide some guidance for those concerned with the management of intellectual property in research tools in molecular biology. Those themes are discussed in the final two sessions, which include a section on different perspectives represented by the workshop participants and a general summary of the workshop.

REFERENCES

Henderson R, Jaffe AB, Trajtenberg M. 1995. Universities as a source of commercial technology—a detailed analysis of university patenting 1965-1988. Working paper No. 5068. National Bureau of Economic Research Inc.

Henderson R, Jaffe AB, Trajtenberg M. 1994. Numbers up, quality down? Trends in university patenting 1965-1992. Paper presented at the CEPR/AAAS conference "University Goals, Institutional Mechanisms, and the 'Industrial Transferability' of Research."

Weiner C. 1989. Patenting and academic research: Historical case studies. In: Weil V and Snapper JW, editors. Owning Scientific and Technical Information. New Brunswick, NJ: Rutgers University Press.

2

Patenting Research Tools and the Law

Rebecca Eisenberg, University of Michigan Law School

Over the past 15 years, a number of legal and commercial developments have converged to make intellectual property issues particularly salient in biomedical research.

A series of judicial and administrative decisions has expanded the categories of patentable subject matter in the life sciences. For many years it appeared that patents on living subject matter would violate the longstanding principle that one may not patent products or phenomena of nature.[1] But in 1980 the US Supreme Court held in the case of *Diamond v. Chakrabarty*[2] that a living, genetically altered organism may qualify for patent protection as a new manufacture or

[1] The US Supreme Court relied on this principle in *Funk Brothers Seed Co. v. Kalo Inoculant Co.,* 333 US 127 (1948), holding invalid a patent on a mixed culture of different strains of bacteria used to inoculate the roots of different species of plants. The court reasoned that, "The qualities of these bacteria, like the heat of the sun, electricity, or the qualities of metals, are part of the storehouse of knowledge of all men. They are manifestations of laws of nature, free to all men and reserved exclusively to none." Although subsequent US cases have never explicitly overruled *Funk Brothers,* in retrospect it seems to represent a high water mark for the "products of nature" doctrine. A long line of lower court decisions has upheld patents on purified forms of products that occur naturally only in an impure state, including purified prostaglandins, *Kuehmsted v. Farbenfabriken,* 179 F. 701 (7th Cir. 1910), *cert. denied,* 220 US 622 (1911), purified aspirin, *In re Bergstrom,* 427 F.2d 1394 (C.C.P.A. 1970), purified adrenaline composition, *Parke-Davis & Co. v. H.K. Mulford & Co.*, 189 F. 95 (S.D.N.Y. 1911), and even a purified bacterial strain, *In re Bergy,* 596 F.2d 952 (C.C.P.A. 1979).

[2] 447 US 303 (1980).

composition of matter under Section 101 of the US Patent Code. Characterizing Chakrabarty's invention as "a new bacterium with markedly different characteristics from any found in nature" and "not nature's handiwork, but his own," the Court indicated that Congress intended the patent laws to cover "anything under the sun that is made by man." With this broad directive from the Supreme Court, the US Patent and Trademark Office (PTO) expanded the categories of living subject matter that it considered eligible for patent protection to include plants[3] and animals.[4]

During the same time period, the explosion of commercial interest in the field, and the concomitant emergence of commercial biotechnology companies, have amplified the importance of intellectual property in the biomedical sciences. Many biotechnology firms have found a market niche somewhere between the fundamental research that typifies the work of university and government laboratories and the end product development that occurs in more established commercial firms. To survive financially in this niche, biotechnology firms need intellectual property rights in discoveries that arise considerably upstream from commercial product markets. This creates pressure to patent discoveries that are closer to the work of research scientists than to ultimate consumer products.

Another contemporaneous development that has contributed to the prevalence of intellectual property in biomedical research is the passage of the Bayh-Dole Act and the Stevenson-Wydler Act in 1980, and a series of subsequent acts that refine those statutes and expand their reach.[5] These statutes encourage research institutions to patent discoveries made in the course of government-sponsored research. For some institutions involved in health-related research, this represented a 180° shift in policy. A generation ago, the prevailing wisdom

[3] In 1985, the PTO held that plants were eligible for standard utility patents, and not merely the more limited rights provided under special statutes for the protection of plant varieties. *Ex parte Hibberd*, 227 USPQ2d (BNA) 443 (Pat. Off. Bd. App. 1985).

[4] The PTO held that oysters were patentable subject matter in *Ex parte Allen*, 2 USPQ2d (BNA) 1425 (Bd. Pat. App. & Int. 1987). Shortly thereafter, the Commissioner of Patents issued a notice stating that the PTO would consider non-naturally occurring, non-human, multicellular living organisms—including animals—to be patentable subject matter. US PTO, Commissioner's Notice, Official Gazette of the Patent and Trademark Office 1077:24 (April 21, 1987). The notice hastened to add that PTO would not consider human beings to be patentable subject matter, citing restrictions on property rights in human beings. The first patent on a genetically altered animal was issued in April of 1988 to Harvard University for the development of a mouse bearing a human oncogene. US Patent No. 4,736,866 (April 12, 1988). The decision to extend patent protection to animals generated considerable public controversy and has been the focus of numerous hearings in the US Congress. Restrictive legislation has been proposed from time to time, including a moratorium on animal patenting, although no such legislation has been passed. More recently, the patenting of DNA sequences has emerged as a new focal point for aversion to patents in the life sciences, but no new legislation has been enacted.

[5] These statutes are codified as amended at 35 USC §§ 200-211, 301-307 and at 15 USC §§ 3701-3714.

was that the best way to assure full utilization of publicly-sponsored research results for the public good was to make them freely available to the public. Today, federal policy reflects the opposite assumption. The current belief is that if research results are made widely available to anyone who wants them, they will languish in government and university archives, unable to generate commercial interest in picking up where the government leaves off and using the results to develop commercial products. To make government-sponsored research discoveries attractive candidates for commercial development, institutions performing the research are encouraged to obtain patents and to offer licenses to the private sector. As a result, institutions that perform fundamental research have an incentive to patent the sorts of early stage discoveries that in an earlier era would have been dedicated to the public domain. A big part of the resulting increase in patenting activity among public sector research institutions has been in the life sciences.

Taken together, these factors have created a research environment in which early stage discoveries are increasingly likely to be patented, and access to patented discoveries is increasingly likely to be significant to the ongoing work of research laboratories.

PATENTS AS A STRATEGY FOR PROTECTION OF INTELLECTUAL PROPERTY

In order to assess the significance of these developments, it is necessary to understand something about patents and their relationship to other forms of intellectual property protection. The term intellectual property is used to refer to a wide range of rights associated with inventions, discoveries, writings, product designs, and other creative works. Some of these rights, such as patents, have more of the attributes of property than others, such as trade secrets. Some of these rights, such as patents, are protected under federal law, while others, such as trade secrets, are a matter of state law. A patent confers a right to exclude anyone else from using an invention, even an innocent infringer who independently develops the same invention without any knowledge of the patent holder's rights. Trade secrets, in contrast, receive more limited protection and may not be enforced against innocent infringers. Although trade secret rights are weaker than patent rights, the availability of legal protection for trade secrets under state law provides an alternative to protection that some inventors might choose in situations where patent protection is unavailable. One therefore cannot assume that withholding patent protection from research tools will improve their availability. Given commercial interest in the development and dissemination of research tools, in the absence of patent protection, firms may be more likely to resort to trade secrecy than to dedicate their research tools to the public domain, which could aggravate the problem rather than resolve it.

Trade secrecy is one way to keep inventions and discoveries out of the hands of competitors to protect an investment in R&D. As long as no one else knows the Coca-Cola formula, the Coca-Cola Company does not have to worry about competition from outsiders who did not share the cost of developing it. But secrecy only works for inventions that can be exploited commercially without disclosure, such as manufacturing processes. Many inventions and discoveries are self-disclosing once they are put on the market in the form of a product, and thereafter may only be protected through a patent. Even when secrecy is feasible, it might not be desirable. From the perspective of an innovating firm, disclosure of underlying technology might help in the marketing of a new product, and from a broader social perspective, secrecy might impede further technological progress in the field.

For some inventions, patents provide an alternative strategy for protecting intellectual property rights that does not require (and indeed does not permit) secrecy. To get a patent, it is necessary to file an application that includes a full disclosure of the invention and describes how to make and use it. In many parts of the world, this disclosure will be made public 18 months after the filing date of the application. In the US, the disclosure is made public when the patent is issued. Once the patent application is on file, disclosure will not jeopardize the applicant's prospects for obtaining a patent.

A patent gives an inventor the rights to exclude others from making, using and selling the invention for a limited term, 20 years from the application filing date in most of the world. During the patent term the inventor may choose to make, use, and sell the patented invention herself (assuming this does not violate the patent rights of others or any applicable laws), or to license others to do so on an exclusive or non-exclusive basis, or even to suppress the use of the invention entirely. One thing an inventor who wants a patent cannot do is keep the invention secret.

BENEFITS AND COSTS OF PATENTS

In the industrial realm, patents are generally believed to promote technological progress in two ways: by providing an economic incentive to make new inventions and to develop them into commercial products and by promoting disclosure of new inventions to the public. The extent to which the patent system achieves these goals is essentially an empirical question with different answers in different fields. In the absence of a patent system, it is unlikely that invention and technical disclosure would come to a standstill. Firms that introduce new technologies into the market would surely find some R&D profitable even without patent rights. The anticipated advantages of being the first firm in the market with an innovation might be enough to motivate some firms to continue investing in R&D. But at least in some fields, the prospect of obtaining patent rights undoubtedly increases incentives to invest in R&D and to disclose research results somewhat.

Patent systems also entail social costs that need to be weighed against the benefits. The most obvious of these social costs is that patents create monopolies that increase the prices and reduce the supplies of the products they cover. This may be a tolerable cost for socially useful inventions that would not be made without the incentives of the patent system (presumably we would prefer to have these inventions at a high price than not to have them at all), but it is a very high cost for inventions that would be developed even in the absence of patent rights. Conferring exclusive rights on inventions that would be made without the incentive of patents reduces the use and increases the price of these inventions without furthering technological progress. Patent systems therefore use rules of law that attempt the difficult task of distinguishing between inventions that would occur even without patents and inventions that require the incentive of a patent. These legal rules call for a comparison between the invention and the "prior art," or preexisting knowledge in the field.

REQUIREMENTS FOR PATENT PROTECTION

The basic requirements for patent protection are novelty, utility, and nonobviousness. *Novelty* means that the invention did not exist before. Determining whether an invention is new requires searching through certain categories of prior art to determine the state of knowledge in the field at the time that the invention was made. Sources of prior art include prior patents, publications, and inventions that were previously in public use. If an invention was already known or used before the time that the inventor claims to have made it, the public gains nothing by conferring a patent. The patent will take something away from the public that it was previously free to use without in any way enriching the public storehouse of knowledge.

The prior art is also relevant to the standard of *nonobviousness*. This standard asks whether the invention constitutes a significant enough advance over what was known previously to justify patent protection. Under US law, the requirement is satisfied if, at the time the invention was made, it would not have been obvious to a person of ordinary skill in the field and who was knowledgeable about the prior art. This determination looks to the level of inventive skill of others working in the field, as well as the state of the prior art. In principle, the requirement might be justified as a means of distinguishing between trivial inventions that require no special incentive to call forth, and more elusive (and, perhaps, more costly) inventions that might not be developed without the enhanced assurance of profitability that patent protection offers. But how the standard will apply in any given case is often difficult to predict, and this uncertainty reduces the value of patents.

The *utility* requirement limits patent protection to inventions with practical applications, as opposed to basic knowledge. The meaning of this requirement has varied over the years from a minimal standard that the invention not be positively harmful to people to a stricter requirement in recent years of safety and effectiveness that has sometimes approached what the FDA would require for approval of a new drug. Recent developments in the courts and in the PTO suggest that the utility requirement may be receding from its recent all-time high level as an obstacle to patent protection. The conceptual underpinnings of the utility requirement are not always clear, but in theory it can be justified as a means of distinguishing between basic research discoveries that are more likely to be effectively utilized if left in the public domain and more practical technological applications that may require a patent to ensure adequate incentives for commercial development. The Supreme Court has stated that discoveries whose only value is as an object of scientific inquiry do not satisfy the utility standard, suggesting that utility could be an important limitation on the use of the patent system to protect research tools.

EXPERIMENTAL USE EXEMPTION

In some cases, the courts have recognized what has come to be known as an experimental use exemption, or research exemption, from infringement liability. On its face, the patent statute does not appear to permit any unlicensed use of a patented invention, in research or otherwise, but language in some judicial opinions nonetheless suggests that use of a patented invention solely for research or experimentation is, in principle, exempt from infringement liability.

The experimental-use doctrine was first expounded in 1813 by Justice Story in *dictum*[6] in the case of *Whittemore v. Cutter*.[7] He observed "that it could never have been the intention of the legislature to punish a man who constructed [a patented] machine merely for philosophical experiments or for the purpose of ascertaining the sufficiency of the machine to produce its described effects."

It is difficult to discern the scope of this exception with any precision, inasmuch as experimental use becomes an issue only in patent infringement actions, and patent holders are unlikely to file a lawsuit against an academic researcher whose use of the invention is commercially insignificant. Judicial pronouncements on the scope of the experimental use exemption address situations in which a patent holder has found a defendant's activities sufficiently annoying to be worth the trouble of pursuing a lawsuit; this factor has undoubtedly skewed the distribution of cases in which the defense arises toward cases with high commercial stakes. Within this universe, the experimental use defense has been frequently raised, but almost never sustained. Nonetheless, courts have consistently

[6] The legal term *dictum* refers to something said in a judicial opinion that was not necessary to resolve the case before the court, and therefore does not create binding precedent in subsequent cases.

[7] 29 F. Cas. 1120 (C.C.D. Mass. 1813) (No. 17,600).

recognized the existence of an experimental use defense in theory, although the defense has almost never succeeded in practice.

Recent case law suggests that the experimental use defense may be available only for pure research with no commercial implications, if such a thing exists. In *Roche Products v. Bolar Pharmaceutical Company*,[8] a 1984 decision of the US Court of Appeals for the Federal Circuit,[9] the court rejected the arguments of a generic drug manufacturer that the experimental use defense should apply to its use of a patented drug to conduct clinical trials during the patent term. The purpose of the trials was to gather data necessary to obtain FDA approval to market a generic version of the drug as soon as the patent expired. The court characterized the defense as "truly narrow," noting that the defendant's use of the drug was "no dilettante affair such as Justice Story envisioned." The court held that the defense does not permit unlicensed experiments conducted with a view to the adoption of a patented invention for use in an experimenter's business, as opposed to experiments conducted for amusement, to satisfy idle curiosity, or for strictly philosophical inquiry. Although it is not entirely clear what sort of research the court would exclude from infringement liability as a mere "dilettante affair," the language of the decision offers little hope of an exemption for research scientists who use patented inventions with an aim to discover something of potential usefulness. It certainly suggests that the defense would be unavailable whenever the defendant's research is motivated by a commercial purpose. As a practical matter, this parsimonious approach could seriously limit the availability of the defense in fields of research with commercial significance, in which even academic researchers are often motivated, at least in part, by commercial interests.[10]

Congress has partially abrogated the decision of the Federal Circuit in *Roche v. Bolar* in the specific context of clinical trials of patented drugs by an amendment to the patent statute.[11] As amended, the statute explicitly permits the use of patented inventions for the purpose of developing and submitting information under laws regulating the manufacture, use, or sale of drugs. But the amendment did not address the broader question of when the experimental use defense would be available outside of that very narrow setting.

---

[8] 733 F.2d 858 (Fed. Cir.), *cert. denied*, 469 US 856 (1984).

[9] The Court of Appeals for the Federal Circuit was created by Congress in 1982 in an effort to create greater uniformity in patent law. It has appellate jurisdiction over decisions of the PTO and of the federal district courts in matters of patent law. Although its decisions may be appealed to the US Supreme Court, that court rarely grants review of its decisions.

[10] For example, the Bayh-Dole Act in effect directs academic institutions to be alert to potential commercial implications of their research so that they can obtain patents as appropriate. See 35 USC §§ 200-212.

[11] Drug Price Competition and Patent Term Restoration Act of 1984, Public Law 98-417, codified in pertinent part at 35 USC § 271(e).

Other countries have more broadly available experimental use defenses than the US, often explicitly included in the text of foreign patent statutes. But even these defenses typically distinguish between experimenting *on* a patented invention—that is, using it to study its underlying technology and invent around the patent, which is what the exemption covers—and experimenting *with* a patented invention to study something else, which is not covered by the exemption. In other words, even outside the US, the defense is not available for researchers who make use of patented research tools in their own work, as opposed to those who study the research tools themselves.

It is difficult to imagine how a broader experimental use defense could be formulated that would exempt the use of research tools from infringement liability without effectively eviscerating the value of patents on research tools. The problem is that researchers are ordinary consumers of patented research tools, and that if these consumers were exempt from infringement liability, patent holders would have nowhere else to turn to collect patent royalties. Another way of looking at the problem is that one firm's research tool may be another firm's end product. This is particularly likely in contemporary molecular biology, in which research is big business and there is money to be made by developing and marketing research tools for use by other firms. An excessively broad research exemption could eliminate incentives for private firms to develop and disseminate new research tools, which could on balance do more harm than good to the research enterprise.

RESEARCH TOOLS IN MOLECULAR BIOLOGY

Molecular biology provides a useful focal point for examining the effect of intellectual property on the dissemination of research tools. It is a dynamic and productive field of research that provides a wealth of new discoveries that are simultaneously inputs into further research and also candidates for commercial development. The obvious implications of discoveries in molecular biology for human health raise the stakes of striking the right balance between public access and private property, particularly when public attention is riveted upon the rising costs of health care. And it profoundly affects the interests of two different types of commercial firms—young biotechnology firms and large, integrated pharmaceutical firms—both of which are sensitive to intellectual property but for different reasons.

This dichotomy between biotechnology firms and pharmaceutical firms oversimplifies the wide range of firms with interests in molecular biology, but it is nonetheless a useful heuristic assumption to help sort through the interests of different sorts of firms. Young biotechnology firms typically need to raise funds to keep their research operations moving forward before they have products to sell to consumers. For these firms, an intellectual property portfolio might be critical at an early stage in their R&D to give them something to show investors as evidence of their potential for earning high returns in the future. With this purpose in mind, they are likely to seek patents on discoveries that are several stages removed from a final product that is ready to be sold to consumers.

Established pharmaceutical firms are also very sensitive to intellectual property rights, but for different reasons and at a different stage in the R&D process. Pharmaceutical firms do not need to go to the capital markets to fund their research; they typically fund new research projects out of profits on existing products. For these firms, intellectual property is not a means of raising capital, but simply a means of ensuring an effective commercial monopoly for their products. A monopoly position in a new drug will help them recoup what might amount to hundreds of millions of dollars required for FDA-mandated clinical testing before they can bring that drug to market. For this purpose, they seek patent rights that cover the downstream products that they sell to consumers, not the upstream discoveries that they may use along the road to product development.

Since they have different reasons for requiring intellectual property rights, these different types of firms are likely to be affected differently by different legal rules. We need to keep the interests of both of these types of firms in mind, along with the interests of researchers and the institutions that fund research, as we think about how to manage intellectual property rights in research tools. Strategies that work for some players could be disastrous for others.

PATENTS ON RESEARCH TOOLS

"Research tools" is not a term of art in patent law. No legal consequences flow from designating a particular discovery as a research tool. Research tools are not categorically excluded from patent protection (except insofar as they lack patentable utility), nor is the use of patented inventions in research categorically exempted from infringement liability.

Nonetheless, there are reasons to be wary of patents on research tools. Although the ultimate social value of research tools is often difficult to measure in advance, it is likely to be greatest when they are widely available to all researchers who can use them. For years, we have sustained a flourishing biomedical research enterprise in which investigators have drawn heavily upon discoveries that their predecessors left in the public domain. Yet the nature of patents is that they restrict access to inventions to increase profits to patent holders. An important research project might call for access to many research tools, and the costs and administrative burden could mount quickly if it were necessary for researchers to obtain separate licenses for each of these tools.

The effects of patenting research tools will vary. For example, patents are unlikely to interfere substantially with access to such research tools as chemical reagents that are readily available on the market at reasonable prices from patent holders or licensees. Many of the tools of contemporary molecular biology research are available through catalogs under conditions that approach an anonymous market. Some are patented, but the patents are unlikely to interfere with dissemination. Indeed, it might be cheaper and easier for researchers to obtain such a tool from the patent holder or from a licensed source than it is to infringe the patent by making it themselves. But not all research tools are of that character.

Some research tools can only be obtained by approaching the patent holder directly and negotiating for licenses; in this context, patents potentially pose a far greater threat to the work of later researchers. Negotiating for access to research tools might present particularly difficult problems for would-be licensees who do not want to disclose the directions of their research in its early stages by requesting licenses. Another risk is that the holders of patents on research tools will choose to license them on an exclusive basis rather than on a nonexclusive basis; this could choke off the R&D of other firms before it gets off the ground. Such a licensing strategy might make sense for a startup company that is short on current revenues, even if it does not maximize value in the long run from a broader social perspective.

Another risk is that patent holders will use a device employed by some biotechnology firms of offering licenses that impose "reach-through" royalties on sales of products that are developed in part through use of licensed research tools, even if the patented inventions are not themselves incorporated into the final products. So far, patent holders have had limited success with reach-through royalty licenses. Firms have been willing to accept a reach-through royalty obligation for licenses under the Cohen-Boyer patents on basic recombinant DNA techniques, perhaps because those patents include broad claim language that covers products developed through the use of the patented technology. But reach-through royalties have met greater market resistance for other patents, including the patents on the Harvard onco-mouse and the polymerase chain reaction (PCR).

Licenses with reach-through royalty provisions might appear to solve the problem of placing a value on a research tool before the outcome of the research is known. One difficulty in licensing research tools is that the value of the license cannot be known in advance, so it is difficult to figure out mutually agreeable license terms. A reach-through royalty might seem like a solution to this problem, in that it imposes an obligation to share the fruits of successful research without adding to the costs of unsuccessful research. But it takes little imagination to foresee the disincentives to product development that could arise from a proliferation of reach-through royalties. Each reach-through royalty obligation becomes a prospective tax on sales of a new product, and the more research tools are used in developing a product, the higher the tax burden.

A further complication arises in the case of inventions that have substantial current value as research tools but might also be incorporated into commercial products in the future. It might be necessary to offer exclusive rights in the ultimate commercial products to innovating firms to give them adequate incentives to develop the products, but it might be impossible to preserve this option without limiting dissemination of the inventions for their present use as research tools.

For all of these reasons, exclusive rights risk inhibiting the optimal use of research tools and interfering with downstream incentives for product development. Much depends on whether the holders of exclusive rights can figure out how to disseminate research tools broadly without undermining their value as intellectual property.

These are difficult problems that defy facile solutions. One of the purposes of this workshop is to examine the solutions that different institutions have come up with and see how they have operated in practice. Which mechanisms have worked well, which have worked badly, and what can we learn from the experiences of others? We need to keep in mind that this issue implicates the interests of many different players who value intellectual property in different ways and for different purposes.

3

Economic Theories About the Costs and Benefits of Patents

Richard D. Nelson and Roberto Mazzoleni, Columbia University

This paper provides a broad overview of theories about the principal costs and benefits of patents and discusses assumptions about the contexts in which inventions are made or developed. Consideration of different contexts suggests that patents play different roles in different technologies and sectors. In some contexts, several patent theories have a degree of plausibility; in others, none of them is very plausible. The question is under which conditions the theories make sense and under which they do not.

In 1958, Fritz Machlup reviewed how economists view the patent system. He reported that economists tended to be negative about the value of the patent system to society, reflecting their concern that patents generate monopolies and that, in many cases, patents are not even necessary to encourage invention. His own position, however, was that there were no good models to replace the patent system and that it serves some useful purposes.

In this paper we identify four broad theories about the principal purposes of patents:

- *Invention-inducement Theory:* The anticipation of receiving patents provides motivation for useful invention.
- *Disclosure Theory*: Patents facilitate wide knowledge about and use of inventions by inducing inventors to disclose their inventions when otherwise they would rely on secrecy.
- *Development and Commercialization Theory:* Patents induce the investment needed to develop and commercialize inventions.

- *Prospect Development Theory*: Patents enable the orderly exploration of broad prospects for derivative inventions.

The four theories are not necessarily mutually exclusive. The anticipation of a patent might stimulate an invention, and the holding of a patent might stimulate its subsequent development. But some versions of the theories are at odds. For example, one version of the disclosure theory assumes that inventions will occur without patents and that the existence of patents widens their use. That is quite the opposite of the most familiar version of the invention-inducement theory, which assumes that invention is motivated by the anticipation of patents.

In general, however, the theories differ in the assumptions that they make about the conditions under which inventions are made, developed, or commercialized. The assumptions are made about the following conditions:

- The nature and effectiveness of means other than patents to induce invention and related activities.
- The likelihood of a group of potential inventors to work on diverse and noncompeting ideas or to be focused on a single alternative or a set of closely connected ones.
- The transaction costs of licensing an invention with and without patents.
- Whether the multiple steps in the invention, development, and commercialization of a new technology tend to proceed within a single organization or several organizations tend to be involved at different stages of the process.
- The topography of technological advance—how inventions are linked to each other both temporally and as systems.

We will examine those assumptions to show the strengths and weaknesses of all the theories.

INVENTION-INDUCEMENT THEORY

The theory that patents motivate useful invention is the most familiar theory of the benefits of patenting. Indeed, much discussion about the benefits of patents proceeds as though motivating useful invention were the only social purpose served by patents and patents always serve this purpose. In fact, as explained later, the situation can be much more complicated in many cases.

All versions of the invention-inducement theory presume either that if there is no patent protection there will be no invention or, more generally, that without a patent system incentives for invention will be too weak to reflect the public interest. In particular, they assume that stronger patent protection will increase the amount of invention.

Under what might be called the canonical versions of the invention-inducement theory—versions associated with the models of economists Arrow (1962), Nordhaus (1969), and Scherer (1972)—inventors, as a group, are implicitly assumed to be diverse, working on different and generally noncompeting things. Thus in the absence of redundant efforts that might occur if many groups worked on competing things, stronger patent protection results in a greater number of useful inventions.

Most articulations of the invention-inducement theory presume that an invention is used or sold by the firm that made the invention. However, Arrow (1962) and more recently Merges (1995) and Arora and others (1994) address the problem that an inventor has in selling an invention to someone else in the absence of legal property rights to it and have taken the position that strong property rights to an invention reduce the transaction costs of licensing it. Strong patents would then also serve the purpose of providing incentives to invent for parties who are limited in the extent to which they can use the invention themselves, by facilitating the sale of rights to an invention.

In most versions of the invention-inducement theory it is assumed, generally implicitly, that the social benefit of a particular invention is strictly its final use value; the social benefit of patent protection stems, therefore, from the additional invention induced by the prospect of a patent. And the social cost of a patent is the restriction on the use associated with the monopoly power lent by a patent. That formulation of the invention-inducement theory leads naturally to analysis of optimal patent "strength," defined as duration (Nordhaus 1969; Scherer 1972), or breadth (Klemperer 1990), and the tradeoff between the amount of increased invention induced by greater patent strength and the increased costs to society associated with the stronger monopoly position of the patent holder (see also Gilbert and Shapiro 1990).

The issues of the consequences of greater patent duration or scope are more complicated if an invention is not only useful as is, but also provides the basis for second-generation inventions. Arrow especially called attention to the possibility that the principal use of some inventions is as input for further inventions (such as PCR, see Chapter 5). Van Dijk (1994) considers what he calls the height of a patent, by which he means the extent to which the patent controls later improvements and variations in the initial invention. Those considerations lead us to the development and commercialization theory, and prospect-development theory, which are discussed later.

We stress that the version of the invention-inducement theory that we have been considering up to now presumes that more inventive effort and more inventors mean more useful inventing. The theory takes on a different look if, instead, all inventors are assumed to be focused on the same set of paths to invention. This assumption gives rise to the "patent race models" of Loury (1979) and Dasgupta and Stiglitz (1980a) and, if the assumed common focus is on a broader but still limited "pool" of invention prospects, to the "overfishing" models of Barzel (1968) and Dasgupta and Stiglitz (1980b). Under either model, patents no longer provide an unambiguous benefit when there are increases in the total inventive effort exerted at any one time or in the number of persons engaged in inventive activity. If inventors perceive that other inventors are in the game, they will see that their returns depend not simply on whether they achieve an invention, but on whether they achieve it first. That might induce them to invest their resources faster or more widely than would be appropriate if the objective were defined simply in terms of achieving a particular invention most efficiently. If inventors are unaware of or ignore the presence of other inventors in the game and calculate their expected benefits as though they had no competition, then from the vantage point of standard welfare economics there will be "too many" inventors playing the game. Under either model, patents will constitute a winner-take-all system. The lure of a patent, therefore, induces inefficient inventive effort in a competitive context. Another possibility suggested by Dasgupta and Stiglitz (1980a) is that the recognition that others are likely to be running toward a particular objective will deter parties from engaging in inventive work in a given field.

It seems that a consequence of that kind of invention inefficiency induced by strong patents would be to shift the tradeoff between the benefits and costs of stronger patents so as to increase the costs. Thus, society ought to opt for stronger patents in fields in which stronger intellectual property protection yields a larger flow of valuable inventions, rather than in fields in which stronger patents lead largely to more hounds barking up the same tree. And that is the case in the model of optimal patent duration developed by McFetridge and Rafiquzzaman (1986).

In any case, under the invention-inducement theory, the basic presumption is that if the award of a patent is not necessary to induce an invention, then it is not in the social interest to offer or grant a patent. That raises the question, "How important is the anticipation of a patent in the inducement of inventing?"

The best empirical evidence suggests that among firms engaged in R&D, patents are an important part of the inducement for invention in only a small number of industries (for example, see Levin and others 1987). Mansfield (1986) asked chief R&D executives of 1,000 US manufacturing firms to identify the fraction of inventions developed by their firms between 1981 and 1983 that they would not have chosen to develop if they had been unable to obtain patent protection. For electrical equipment, primary metals, instruments, office equipment, motor vehicles, and several other industries, the fraction would have been less than 10%. Executives in those companies rated what they could gain by patents as much less important than the advantages that go with a head start in reaping returns from their inventions. Pharmaceuticals and fine chemicals were important exceptions: executives in the pharmaceutical industry reported that without patent protection 60% of their new pharmaceuticals would not have been developed, and the reduction in "other chemicals" would have been about 40%.

However, the studies mentioned above focused on large and medium firms with R&D laboratories, which are for the most part in a position to exploit their inventions by using them themselves. They do not depend on sale or license of their inventions for returns. For inventors who must sell or license to reap returns, patents might be far more important. It appears that in some sectors patents are an essential part of the inducement for inventing.

There is a proliferation of empirical work on what patents are about, who uses them, and how important they are. This work is motivated entirely by the invention-inducement theory; that is why a lot of what was learned in the work might not be relevant to research tools.

The case can certainly be made that for many university inventions that were funded with public money, the policy implication of the invention-inducement theory is that a patent should not be granted if it is not necessary to grant a patent to get an invention. The results of research would be published in any case. In many instances, firms will have ample incentive to work with and develop what comes out of university research. They can usually patent their developments, gain the advantage of a head start in the market, or both. No grant of an extensive license based on anticipated economic changes is needed to motivate this work. Moreover, the presence of a patented invention, with a requirement for would-be developers to get a license to do further work on the original idea, restricts the number of parties that will engage in that work. The latter argument against patenting seems particularly strong if potential developers are diverse in the directions that they might follow and if the licensing arrangement of preliminary ideas, whose ultimate commercial value is unclear, are not easy to work out.

DISCLOSURE THEORY

The primary issue raised by the disclosure theory is not so much whether strong patents encourage more inventing, but rather how inventors reap the returns from their inventions. It presumes that secrecy is possible and sufficient to induce invention but that society is better off granting intellectual property rights and getting disclosure in exchange. A patented invention would thus be available for uses that the inventor did not know about or was not in a position to implement. Under this theory, a patent both advertises the presence of an invention and facilitates licensing. That argument, in effect, turns the invention-inducement theory on its head: patents are not necessary to induce invention, but rather what patents do is encourage disclosure and, given some assumptions about the transaction costs of licensing the invention, enable it to be used more widely than it would be without a patent.

Figuring out what makes a newly-marketed product work is substantially easier than the initial invention and development of that product. Thus, as suggested by Levin and others (1987), secrecy would seem more effective for process than for product inventions. From that perspective, the most relevant domain of the disclosure theory might be process inventions. Various studies have shown that in some industries, firms customarily engage in general cross licensing of their process technologies, a sharing of technology that might not occur if patents on processes were not available.

However, a broader view of the disclosure theory opens the door to a wider possible range of applicability. The possession of a patent, rather than simple reliance on nonpatent measures to reap returns, might make a firm more willing to advertise its inventions and to contract to give technological information and assistance to a noncompeting firm to help it to adapt the invention to its own uses. This variant of the theory is a close kin to the version of the invention-inducement theory in which inventors have limited ability to exploit an invention themselves. Under this variant, the possession of a patent might make the firm more receptive to proposals by firms in other lines of business to develop the invention for different uses. Such possibilities lead to the development and commercialization theory and the prospect-development theory.

DEVELOPMENT AND COMMERCIALIZATION THEORY

In its simplest version, the theory that patents induce the development and commercialization of inventions seems to be a variant of the invention-inducement theory, but with patenting occurring early in the process of inventing and with much additional work needed before the crude "invention" is ready for actual use. A patent at an early stage is seen as providing the assurance that if the development is technologically successful, its economic rewards will be capturable, thus inducing a decision to develop it.

Rebecca Eisenberg has called our attention to a supplement to this argument: that the possession of a patent enables the patent holder to go to capital markets for development financing. That capability might be important for a small firm faced with large development costs before it can get its invention to market.

The development and commercialization theory is different from the invention-inducement theory for circumstances in which one organization does the early inventing work but is not in a position to do the development work. The original inventor's possession of a patent then facilitates handing off the task to an organization better situated for development and commercialization. Years ago, Willard Mueller (1962) pointed out that many of DuPont's product innovations were based on inventions bought from smaller firms. Similarly, in the 1920s, General Electric bought and developed many inventions made by private inventors or small firms (see Reich 1985).

The development and commercialization theory was widely cited in the discussions that led to the Bayh-Dole act, which gave universities the patent rights on inventions that emanated from their government-funded research projects. The proposition was that although the inventions had been achieved with public funding, they would serve no economic purpose until they were developed to a point where they were commercial, and only companies were able to undertake such development. That constituted a separation of the site of invention from the site of development. Under the version of the development and commercialization theory most clearly articulated in the discussion that led to the Bayh-Dole act, a company would be unlikely to engage in development of a university invention unless it held proprietary rights. If universities held strong patent rights, they would be in a position to sell exclusive licenses. In contrast, if there were no patents, or if the government held them with a commitment to nonexclusive licensing, companies would be unlikely to invest in the necessary development work.

Another interpretation of the development and commercialization theory is that the possession of a patent gives the original patent holder—a university or small firm—an incentive to push its inventions out to firms that can develop and commercialize them. That is basically an extension of the version of the disclosure theory discussed above. It is a different view of the development and commercialization theory from the one that implicitly assumes that without a strong initial patent a firm will not undertake the development work necessary to lead to a profitable product or process innovation.

PROSPECT DEVELOPMENT THEORY

A number of years ago, Edmund Kitch proposed a prospect-development theory of the societal benefits of patents. Like the development and commercialization theory, it proposes that the utility of a patent comes after an initial invention is made.

Kitch's theory was that having a broad patent on an initial invention enabled the patent holder to orchestrate development of a technological prospect in various dimensions, whereas development of an initial invention that was freely available to all would be chaotic, duplicative, and wasteful.

The theory that patents enable orderly development of broad technological prospects differs from the development and commercialization theory in suggesting that a wide range of developments or inventions might become possible if the initial invention is available as an input—through either development or modification in different directions. Many university inventions, particularly research tools, are of this sort.

An implicit feature of various versions of the development and commercialization theory is that although important resources and risk-taking might be needed to develop an invention, there is essentially one product at the end of the commercial rainbow. In its most common formulation, the development and commercialization theory makes the assumption that the work based on an initial invention is not patentable or otherwise appropriable. In the prospect-development theory, Kitch (1977) assumed that there is an abundance of appropriable inventions to be made by using the initial invention as input but suggested that it is problematic. That is, many inventors share knowledge and see the same potential inventions, and they know that their competitors also see them, so there is a lot of racing for specific targets of opportunity and general overfishing in the prospect pond. Thus, a broad patent on the initial invention is necessary if the "mining of the prospect" or the "fishing of the pool" is to proceed in a less-wasteful, less-duplicative fashion.

If you reflect on the prospect-development theory, you will immediately recognize that it depends on a view that is almost antithetical to the notion about what makes progress in science—it depends on the view that it is good to have many people doing different types of things because different ones will see different things and different ones will be more skilled than others at doing different types of things. Trying to make orderly or rationed access to innovations is likely to be socially very costly.

Indeed, under this version of the prospect-development theory, there might be very high social costs to granting a broad initial patent that gives monopoly rights to exploration of the prospect. It would reduce the number of diverse inventors who would be induced to work on the prospect by the lure of a patent down the road, inasmuch as their ability to work on that patent would be constrained by their ability to negotiate a license with the holder of the original prospect-defining patent.

This theory suggests that an important issue defining the benefits and costs of granting patents on broad prospects is what is assumed about the market for patent licenses. If one assumes that, in general, the transaction costs of patent licensing are small, then one may take a relatively relaxed view of the costs of granting a large prospect-controlling patent, even when one believes that potential explorers of the prospect have diverse ideas of what they would do. But if one assumes that transaction costs are high, one is less sanguine about this outcome.

The question of transaction costs is particularly important when technological advances within a prospect are strongly connected. Advances in technology can be connected to each other in two ways. First, technological advance can be *cumulative*, in that today's advance lays the basis for tomorrow's. Although the simplest notion of a prospect is that of an initial node with a large number of potential add-on branches, in fact any of the branches can take the form of a long chain. In such a long chain, ability to operate the most advanced version will require the ability to do things that were the subject of earlier inventions. Second, the operative products or processes can form a *system*, in that they incorporate a number of components. Ability to use the most advanced system might require access to a collection of components. Some of the most important technologies have both attributes. For example, aircraft and computers may be called cumulative system technologies. Merges and Nelson (1990) propose that the historical record show that granting broad patents in cumulative-system technologies is often counterproductive (for instance, research tools). Unless licensed easily and widely, the presence of such patents tends to limit the range of potential users who have access to all components of the technology. In a number of instances, the consequence was to make technological advance difficult and costly.

We suggest that in most fields of technology, when one or a very small number of parties essentially control the inventive effort, not much happens—with the possible exception of Bell Laboratories and AT&T in bygone days.

If you look at the history of aircraft or radio in the United States, strong broad patents that were selectively enforced by the patent holder interfered with the development of the technology. The situation was resolved only when a system for relatively low-cost licensing emerged in the industry or the technology in question, as it did after a while with radio, aircraft, semiconductors, and computers.

In any case, particularly in the prospect-development theory but more generally whenever an invention is viewed as contributing to further invention potential, as well as creating a new or improved product or process of immediately final use, it can be asked whether strong patents enhance or hinder technical advance over the long run. The question of how strong a patent should be, or whether a patent should be granted at all, no longer turns on the analysis of a tradeoff between the positive effects of stronger patents on inventing and the restrictions in use of the technology associated with a regime of strong patents, as in the invention-inducement theory. Rather, a good part of the argument is about whether the *long-term* effect of strong patents is to encourage or discourage innovation.

ISSUES IN PATENT REFORM

There is a good deal of argument about what theory, or what version of a theory, is appropriate to which context, but this question often is glossed over. Almost all empirical work on the role of patents has been oriented by invention-inducement theory, and almost all patent policy issues are argued out on the same terms. Many of today's most important patent policy issues seem not to have been adequately viewed through this theoretical lens.

Consider the debate in the late 1970s that led to the Bayh-Dole act. As we noted, the argument that carried the day was that patents were required if inventions, already achieved under federal funding, were to be developed and commercialized. That is a development and commercialization theory argument with a touch of disclosure theory. We have argued that the particular version the development and commercialization theory put forth most vigorously in these debates—that companies would not develop an invention unless they had a patent on it—probably was not widely relevant. But Bayh-Dole undoubtedly has led to a substantial increase in university entrepreneurial activity. Whether that is good or bad is a complicated question. However, the evaluation of Bayh-Dole, like the arguments that lay behind its genesis, must be cast in terms of all four theories.

The same is true of the issue of patenting the codes on gene fragments identified in the human genome project. Here, too, at least in the early days of the discussion, the issue was not that the prospect of a patent was needed to get the research work done; the research work was being funded by government. Rather, it was felt that patents on the coded gene fragments were needed if companies were to be induced to take that information and use it to achieve commercial products. That is, the arguments were those of the development and commercialization theory and perhaps the prospect-development theory.

As matters have turned out, the belief that codes for gene fragments will be patentable has led to the birth of private for-profit firms, whose business is to discover the codes, in anticipation of profiting from licensing to larger companies that would take on the development work. Interestingly, several large pharmaceutical companies have argued that gene fragments should not be patented, but rather that their identified codings should be in the public domain. Their case is that progress from coded gene fragments to useful final products will cost more if gene-fragment codes are patented than if they are in the public domain. They are essentially arguing that the standard version of the development and commercialization theory has it backwards. And several of these companies are supporting research to identify gene codes, on the condition that the information be put in the public domain. The public-policy issues here are very complex.

The issues surrounding Bayh-Dole and gene fragments differ in important ways but they also have important common elements and raise common questions. Perhaps the most basic question that they raise is whether the presence or prospect of patents stimulates or interferes with technical advance in a field. Obviously, it does not under the invention-inducement theory. But under a more complex theory, the answer is not always apparent. The appropriate domain of patents is badly in need of open examination today. The argument that strong intellectual property rights in a field can smother technical progress is, of course, connected to assumptions about several of the context conditions discussed earlier. To understand better whether our current patent policies help or hinder the achievement of our societal objectives, we need to examine those assumptions rigorously.

The following three questions are put forward as possible starting points for inquiry: In what fields of technology are technical advances so strongly connected to one another, either temporally or in a system of use, that effective inventing today requires access to prior inventions? What are the fields of inventing in which progress generally requires the effective interaction of a number of different organizations? Do patents in fact contribute to or hinder the access and cooperation needed for technical advance in such contexts? As indicated earlier, little empirical research has been aimed at this cluster of questions. Our lack of knowledge limits our ability to analyze intelligently the current pressing issues of patent reform.

REFERENCES

Arora A and Gambardella A. 1994. The changing technology of technological change: general and abstract knowledge and the division of innovative labor. Research Policy Sept. 94: 523-532.

Arrow KJ. 1962. Economic welfare and the allocation of resources for invention. In: Nelson RR, editor. The Rate and Direction of Inventive Activity. New York: Princeton University Press.

Barzel Y. 1968. Optimal timing of innovation. Review of Economics and Statistics 50: 348-355.

Dasgupta P and Stiglitz JE. 1980a. Uncertainty: Industrial structure and the speed of R&D. Bell Journal of Economics 11: 1-28.

Dasgupta P and Stiglitz JE. 1980b. Industrial Sstructure and the nature of innovative activity. Economic Journal 90: 266-293.

Gilbert R and Shapiro C. 1990. Optimal patent length and breadth. RAND J Econ 21: 106-112.

Kitch EW. 1977. The nature and function of the patent system. J Law Econ 20:265-290.

Klemperer P. 1990. How broad should the scope of patent protection be? RAND J Econ.: 113-130.

Levin RC, Klevorick AK, Nelson RR, and Winter SG . 1987. Appropriating the returns from industrial research and development. Brookings Papers on Economic Activity 14: 783-820.

Loury GL. 1979. Market structure and innovation. Qtly J Econ. XCIII: 395-410.

Machlup F. 1958. An economic review of the patent system. Washington, U.S. Govt. Printing Office.

Mansfield E. 1986. Patents and innovation. Mgmt Sci 32: 173-181.

McFetridge DG, and Rafiquzzamanm M. 1986. The Scope and duration of the patent right and the nature of research rivalry. Res Law Econ 8:91-120.

Merges R. 1995. Contracting into liability rules: institutions supporting transactions and intellectual property rights, [manuscript], Berkeley: University of California Law School.

Merges R, and Nelson RR. 1990. On the complex economics of patent scope. Columbia Law Review 90(4): 839-916.

Mueller, WF. 1962. The origins of the basic inventions underlying DuPont's major product and process innovations. In: Nelson RR, editor. The rate and direction of inventive activity, NBER, New York: Princeton University Press.

Nordhaus WD. 1969. Invention, growth, and welfare. A theoretical treatment of technological change. Cambridge, MA: MIT Press.

Reich LS. 1985. The making of American industrial research: science and business at GE and Bell, 1876-1926, New York: Cambridge University Press.

Samuelson P, Davis R, Kapor MD, and Reichman JH. 1994. A manifesto concerning the legal protection of computer programs. Columbia Law Rev 94(8): 2308-2431.

Scherer FM. 1972. Nordhaus's theory of optimal patent life: a geometric reinterpretation. Am Econ Rev 62: 422-427.

Van Dijk TWP. 1994. The limits of patent protection. Maastricht, The Netherlands: Universitaire Pers Maastricht.

4

Access to Data and Intellectual Property: Scientific Exchange in Genome Research

Stephen Hilgartner, Cornell University

Much of the discussion regarding intellectual property policy is framed in terms of a fairly simplistic image of a world of academic science that is very open and a commercial world that is closed and secretive. My goal here is to present a somewhat more nuanced picture of scientific exchange and to explore its implications for the subject of this workshop. I will begin by describing and critiquing a traditional sociological approach, rooted in the work of Robert K. Merton, to analyzing scientific exchange. I will then present an alternative framework for understanding how scientists regulate access to data, illustrating its use with examples from studies of the research community involved in genome mapping and sequencing. To conclude I will suggest some implications of this framework for analyzing how intellectual property policy might affect scientific openness.[1]

SOCIOLOGICAL DIMENSIONS OF SCIENTIFIC EXCHANGE

Developing a sociological explanation of the systems of exchange in which scientists participate is a complex task. One way to state the problem is to imagine that a naive observer (see Latour and Woolgar 1979), innocent of any understanding of the practices and behavior of biomolecular scientists, walks into

[1]This presentation is based on research on data access practices conducted with Sherry Brandt-Rauf and on my studies of the genome mapping and sequencing community. It draws heavily on several papers that are published or in press; see Hilgartner and Brandt-Rauf (1994), Hilgartner and Brandt-Rauf (in press [Stanford]). See also Hilgartner (forthcoming [Private Science]).

a molecular genetics laboratory. The observer is confronted by a dazzling variety of biological materials, texts, software, and instruments. A sequencing machine sits on a bench. Racks of clones fill the freezers. Laboratory notebooks, computer printouts, reprints, and draft manuscripts cover the desks. When our observer asks the scientists where all this stuff came from, he is told different things: some of it was purchased in open markets, some of it was produced locally in the laboratory, and some of it came from colleagues. Regarding the eventual disposition of these items, he finds further variety. Most of them are of little interest to anyone outside the laboratory but a few are of intense interest to people in the world outside. Some of the items will be submitted to scientific journals. Some might be included in patent applications. Some might be shared with colleagues. Some might be kept quietly in the laboratory, no one being told about their existence.

How can one develop a sociological explanation of the traffic patterns of resources in and out of scientific laboratories? How can one explain the process that shapes who gets what, when, and under what kinds of terms and conditions? This formulation of the problem of scientific exchange focuses attention on the particular entities, or resources, that are involved in exchanges—entities that received little notice in early sociological work on scientific exchange. The traditional approach to these problems is rooted in the work of Robert K. Merton (1942), who, in a classic paper published in 1942, laid out a theory of the normative structure of the scientific community. Merton argued that a normative commitment to producing knowledge that becomes the common property of the scientific community is one of the defining characteristics of science. Free and open scientific exchange is important because it allows knowledge claims to be extensively tested by a skeptical scientific community. Only the claims that survive a period of intense scrutiny become scientifically certified, valid knowledge—a form of knowledge that is inherently public and communally held.

Building on Mertonian theory, Warren Hagstrom (1965) developed a gift-exchange model of scientific exchange in which individual scientists contribute their findings to the scientific community and in return can expect to receive various forms of recognition. The gift-exchange perspective has been recently applied to molecular genetics by Katherine W. McCain (1991), and a discussion of her paper provides a useful starting point for our analysis. McCain's argument is based on two distinctions. The first distinction is between *research results* and *research-related information*. Research results are what gets published in journals or technical reports and thereby become the communal knowledge of the scientific community. Research-related information is a residual category that includes all kinds of entities that embody information but cannot be published in journals; it includes clones, algorithms, software, and descriptions of techniques that are too detailed to be included in the methods sections of scientific papers.

The second distinction is between *public science* and *private science*. During the early stages of research, the products of research are the private property of the individual scientist; people might share research products with their immediate colleagues but, in general, they keep their results in the laboratory. Private science, in this model, occupies a temporal phase in the research process: research products remain within the realm of private science until publication occurs. When a researcher decides to publish findings or to present them at meetings, research results cross the private-public boundary. Similarly, at that point, whatever research-related information is needed to replicate the results also becomes public science. In the public science phase, norms of openness govern the exchange of research-related information so that scientists can validate results and advance the field. Problems of lack of openness in this model occur mainly when people fail to provide research-related information after publication of research results. McCain found little indication that that was a widespread practice.

For defenders of scientific openness, McCain's conclusion might be welcome. But rather than confidently deciding to adjourn our workshop early, we might consider some of the limitations of the gift-exchange perspective. Most important, the perspective gives little consideration to the processes through which scientists decide what entities to move across the private-public boundary and when. To understand the traffic patterns discussed above, it is necessary to look more closely at how, why, and when the entities produced in scientific laboratories cross the private-public divide. In addition, the gift-exchange perspective is based on a rather sharp distinction between public and private science, which is at variance with the many gradations of "publicness" that actually occur in scientific practice.

DATA-STREAM PERSPECTIVE

To address those kinds of problems, Sherry Brandt-Rauf and I developed an alternative perspective—the data-stream perspective (Hilgartner and Brandt-Rauf 1994). This framework is informed by a variety of recent social studies of science that emphasize scientific practice and culture, and it is especially amenable to actor-network theory (Callon 1995, Latour 1987). The data-stream perspective conceptualizes data not as isolated objects, but as entities that are embedded in evolving streams of scientific production.

It is important at this point to say something about how Brandt-Rauf and I use the term *data*. We define it inclusively as the many different entities that scientists produce and use during the process of research. In this usage, data include a wide variety of materials, instruments, techniques, and written inscriptions. Such a broad definition is needed because scientists in every subfield have their own specialized conceptual categories for classifying the resources that they use and produce. Distinctions among data, findings, and results, between samples and materials, and among techniques, software, and instrumentation can become confusing when they cross subfields because the terms are not used uniformly.

Brandt-Rauf and I therefore group all the entities described above under the term *data,* and we stress the heterogeneity of the category.

Another important point about data is that they are not useful as isolated entities; only when they are connected to a suite of other resources can a scientist use them to accomplish anything. What one needs to perform scientific work is a complex assemblage of resources and techniques. Data are embedded in complex assemblages that weave together many heterogeneous entities. The assemblages are transformed and manipulated as work proceeds, producing evolving streams of products. For example, streams of inscriptions will evolve as the "raw" output of instruments is manipulated mathematically and incorporated into tables, diagrams, and graphs that in turn are explicated and discussed in written texts (see Latour and Woolgar 1979). Streams of materials also result as samples are purified, analyzed, and otherwise processed. The power of scientific research is in the ability of these assemblages to evolve as people produce new entities by reconnecting and recombining many forms of data in many ways.

One of the most important processes that occur as scientific work proceeds is continuing evaluation of the credibility of various pieces of data. At the research front, the assemblages that constitute data-streams contain elements that vary greatly in perceived credibility. Some entities are considered to be well established; people have great confidence in the accuracy of a particular instrument, the reliability of an observation, the robustness of an assay, or the stability of a cell line. Others are of questionable validity. Perceived credibility often fluctuates throughout the process of scientific production (for example, Collins 1985; Collins and Pinch 1993; Knorr-Cetina 1981; Latour 1987; Latour and Woolgar 1979; Lynch 1985). In general, judgments progress toward a definitive resolution. Some data are rejected and others deemed reliable, but the temporal nature of the evaluation process—with the shifting perceptions that often accompany it—adds complexity to data-streams.

The data-stream perspective emphasizes the continuous properties of scientific production: data are conceived as phenomena that are embedded in evolving assemblages, rather than as discrete entities with unshifting boundaries. The point is to shift the level of analysis to the stream as a whole. In trying to explain the kinds of exchanges in which scientists engage, one cannot assume that data arrive on the scene in neatly packaged units that are naturally ready to be disseminated. Each scientific field has its own conventions about what constitutes a publishable paper and what constitutes an interesting result. Within particular communities of researchers, these conventions can be clearly understood and indeed seem obvious to participants. However, conventions are neither identical across fields nor entirely stable. To provide general insights into how data are shared or otherwise exchanged among scientists, one cannot simply assume that the conventions of a single research community constitute the only way to conduct science; instead, the conventions of different research communities become phenomena for social analysts to explain. The data-stream perspective frames the issue in terms of two central questions: What portions of a given data-stream typically are distributed to whom and under what terms and conditions? How are these portions bounded? In other words, how are discrete entities extracted from the continuously evolving streams of scientific production and entered into exchange relations?

TYPES OF TRANSACTIONS: STRATEGIC CONSIDERATIONS

To understand systems of exchange in science, one must consider the different sorts of transactions in which scientists engage. The limits of the gift exchange perspective will also emerge. McCain (1991), for example, emphasizes two kinds of transactions: transactions between an individual laboratory and the entire scientific community (such as publication of a research paper) and transactions between a source and a requestor (as when one laboratory requests a clone from another). However, these are included in a much broader array of transactions that take place in academic science. It is hard to overstate the importance of open publication in academic reward structures, but it must also be remembered that even in academe publication is only one means among many of distributing data. Data are also quietly given to selected colleagues, they are patented, they are transferred when visitors come to the laboratory to learn novel techniques, they are bought and sold, they are privately released to corporate sponsors, and they are retained in the laboratory pending future decisions about their fate (Hilgartner and Brandt-Raub 1994).

It is useful to consider the strategic issues involved in selecting which of those things to do at what time. A key concept here is *competitive edge*. Using the idealized example of some novel techniques, we can see that it is typical in molecular biology for a new technique initially to be difficult—it might need "magic hands." Later, however, it is increasingly routinized and is packaged in standard protocols; if it is extremely successful, it is incorporated into commercial kits and sold on the market (see Fujimura 1987). In any case, the technique typically becomes available to an increasingly wide circle. As that is going on, the competitive edge that a scientist gets from using the technique in the laboratory declines. At the beginning, because the technique is scarce, a researcher might be able to do things that no one else can do, gaining a short-term competitive advantage that can be important. The competitive edge typically declines as the technique or other initially scarce entity is disseminated.

Various strategies can be used to exploit short-term competitive edges strategically. One of them—widely used by academic scientists—is to restrict access and use data to generate more data that will be cashed in later. Another is to use carefully targeted access; for example, the data can be used as a bargaining chip to negotiate with corporate sponsors or others for resources in return for access. Another is to offer widespread access, say, via open publication, to build one's academic reputation. Often, the question is not necessarily whether to provide

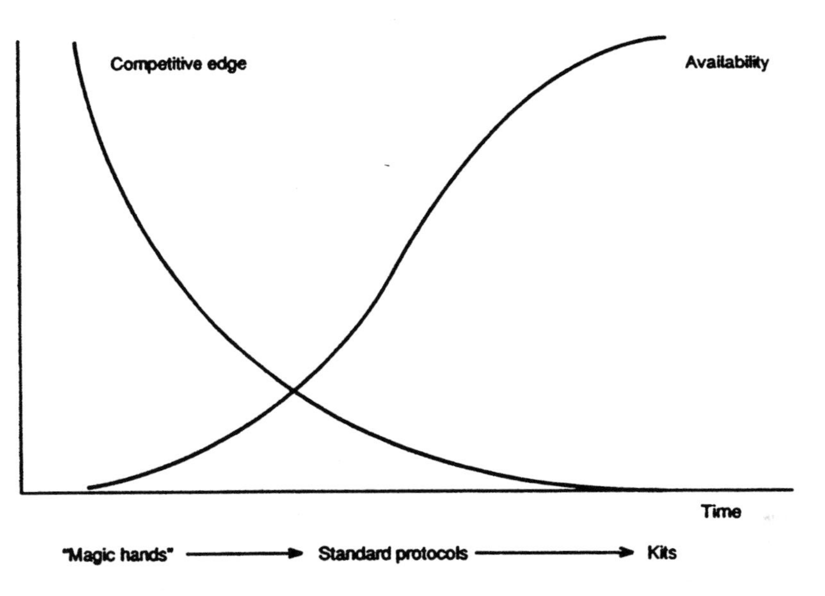

Relationship between the availability of a technique and the competitive edge it conveys. Following initial development, the technique is progressively routinized and becomes available to growing numbers of labs. As this occurs, the comparative advantage it conveys declines. Source: Published with permission from Sage Publications.

access but how much access to provide and when. Here, decision strategies become very complex. Timing can be a key element, both because competitive edge might exist for only a short period of time and because access control interacts in interesting ways with quality control (for discussion in the context of Internet-accessible biomolecular databases, see Hilgartner 1995). Scientists constantly make judgments about whether data are sufficiently reliable to disseminate.

COLLABORATION

Another set of important interactions around the data concerns collaboration (Hilgartner and Brandt-Rauf in press). Scientific collaboration takes many forms, but one important feature of many of them is the merging of portions of data-streams. From this point of view, collaboration involves negotiations in which data serve as bargaining chips in discussions of whether it makes sense to pool resources. Assessing whether or not a collaboration makes sense entails considering the resources that different groups can bring. In addition, because forming a collaboration allows scientists to create a data-stream that spans the boundaries of individual laboratories, negotiations about the terms of the collaboration must typically be conducted. Someone who controls a unique resource is in a position to dictate the terms of all collaborations involving that resource. As one molecular biologist put it to me, "if the other people don't like those terms, they don't have to collaborate." In other situations, complex and protracted negotiations can arise, especially when groups possess relatively equal resources and there is no clear hierarchy governing the relationship (Hilgartner and Brandt-Rauf in press). In some cases, such as the European yeast genome-sequencing program, policy-makers establish formal rules granting entitlements to portions of a data-stream to manage collaboration among geographically dispersed laboratories (Hilgartner and Brandt-Rauf in press).

Even once negotiations are concluded, the resulting collaborations can be fragile and require continuing attention. There seem to be several reasons for that. Many collaborations break down just because the work fails to extend data-streams in the expected directions. Another problem stems from concern that the exchanges among different laboratories have become uneven. A third is ambiguity about what portions of a data-stream are included in a collaboration; this is a problem that is often unavoidable because it might be infeasible to negotiate these matters too tightly in advance. Most of the time, the limits and nature of a collaboration are based on a shared understanding, but shared understandings can easily break down. Social psychologists have shown that in a variety of situations involving collective work (such as basketball games, marriages, and joint writing projects), people tend to perceive their own contribution as being larger than the other participants perceive it, perhaps because it is easier to recall the details of one's own activity than to recollect what other people did (see, for instance, Ross and Sicoly 1982). For all reasons, the negotiations that create and sustain collaborations are not a stage in the research process, but rather a continuing process.

THE DATA-STREAM PERSPECTIVE APPLIED: A SCHEMATIC EXAMPLE

A schematic example might help to illustrate how the data-stream perspective can be used to explain access practices in a particular field of science (from Hilgartner and Brandt-Rauf in press for fuller discussion). Consider the case of hunting for disease genes in the late 1980s or early 1990s, when creating physical maps in the region of a disease gene required a massive amount of work. From high altitude, the data-stream for mapping disease genes can be understood as an evolving assemblage of landmarks in the region where the gene is believed to lie, in which the landmarks become more densely interconnected as work proceeds. Access to such a data-stream could be provided in many ways. At one extreme, a laboratory could use the Internet and Federal Express to provide daily access to its entire data-stream. At the opposite extreme, a laboratory could withhold the entire data-stream and even keep secret the fact that it is looking for the gene.

Those observations suggest a couple of questions, which I will attempt to answer below: What are the typical practices through which gene hunters regulate access to their evolving maps? Can those practices be explained in terms of the structures of data-streams?

Turning first to typical practices, there has been considerable variation within the disease-gene mapping community as to the level of access provided. However, there is no doubt that it takes place only within a portion of the range between the theoretical extremes. In particular, a culture of extreme openness is not to be found: no one provides unrestricted daily access to evolving data-streams in this line of research. Instead, three practices for providing access to data-streams are typical. The first is *nonrelease*, in which scientists hold the data in the laboratory while production proceeds. A second is *delayed release*, in which a large gap is maintained between the time when the data are generated in the laboratory and the time when it is shared outside the laboratory. The third is *isolated release*, in which a laboratory provides access to portions of the data-stream that have been specifically bounded in ways that make the data-stream difficult or impossible to extend.

To explain these practices it is useful to take note of the competitive structure of disease-gene mapping, which was complex and laborious in the late 1980's and early 1990's. Success was by no means guaranteed. Moreover, in the case of the major Mendelian disorders, multiple groups competed to find each gene, and people aptly referred to gene hunts as races. In short, gene hunting was a high-stakes game with a well-defined goal that only one group could reach. It was characterized by intense zero-sum competition.

The second issue to consider is the potential value that different portions of the data-stream have to competitors. In the case of genetic maps that are produced with publicly available clones, published data are of immediate utility to competitors because combining multiple sources of data will usually lead to an increase in the quality of the map in the region. People not only can instantly catch up, but if they are keeping their own data secret at the same time, they might be able to seize the lead. Even if competitors wanted to check the data carefully, the economies are favorable because it takes fewer tests to verify a map than were required to produce it in the first place. It is therefore not surprising that gene hunters do not provide daily releases, and nonrelease can be explained as a consequence of zero-sum competition.

How, then, can we account for delayed release? Because one cannot engage in nonrelease forever! Eventually, it becomes necessary to demonstrate progress to colleagues and funders to sustain a long-term gene hunting effort. Another reason is that scientists use publication as a hedge against the possibility that another group will identify a given gene first, which is fairly likely in many of these races. Although a paper titled "Mapping in the Region of the Gene for Deadly Disease" will be minor in comparison with a paper titled "Positional Cloning of the Gene for Deadly Disease," it has some value. The practice of delayed release thus can be viewed as a strategy for managing a tradeoff between the advantages of publication and the risks of competitors' taking the lead. If the delay is managed carefully, the risks will be minimal.

If we turn to isolated release, the strategic incentives are similar to those for delayed release. However, the practices that enable people to make data-streams impossible to extend are often controversial. A good example of a means of data isolation in the arena of disease-gene mapping is the renaming of clones: by changing the name of a publicly available clone, one can make it unrecognizable, and maps that use the novel name cannot be extended. Renaming clones is considered inappropriate by many researchers, so doing so entails risks to reputation. For some combination of moral and strategic reasons, many disease-gene mappers refrain from such practices, although there are exceptions.

This stripped-down discussion shows how a strategic analysis that looks at the empirical structure of data-streams can go a long way toward explaining scientists' practices regarding access to data. However, this kind of analysis has important limitations. First, at this level of generality, the analysis applies to disease-gene mapping in general and does not take into account the differences in the histories and personalities associated with different genes, chromosomes, and so on. Second, this kind of strategic analysis does not include any discussion of the rhetoric that is used in access negotiations, which can be important in shaping outcomes. Third, the focus on strategic incentives clearly needs to be broadened to include collective definitions of appropriate conduct in science and how those definitions are applied to new situations and renegotiated during their application.

Despite those limitations, the example provides a sense of how scientific exchange can be analyzed from the perspective of data-streams. In addition, it suggests how different this kind of analysis is from the kind that results when one assumes that academic science is governed simply by a culture of openness.

INTELLECTUAL PROPERTY AND OPENNESS

I have argued that rather than merely assuming that academic science is governed by openness, analysts should try to understand the processes that shape what gets made public, what is kept private, and what is deployed in transactions that fall between these extremes. I now want to consider the implications of the more nuanced picture of scientific exchange that I advocate for issues of intellectual property protection. In particular, I want to explore the question of whether we should expect an emphasis on intellectual property in academic science to cause a reduction in scientific openness.

To look at that question, one clearly needs to consider how intellectual property considerations influence a number of aspects of scientists' practices. Do intellectual property considerations influence what portions of data-streams are provided, to whom, and when? Do they introduce new sources of delay? Do they change the kinds of restrictions that are placed on the use of data? Do they increase the complexity and formality of negotiations over access to data? Do they make collaboration more unstable or difficult to form? Do they complicate the development and maintenance of shared understandings about control over data-streams that are collectively produced? Without a doubt, the answers to those questions will vary with the particular context, and many case studies will be needed before these issues are fully understood. However, I want to conclude with several assertions about the likely effects of intellectual property on openness in academic science.

First, there is little reason to believe that intellectual property protection is likely to lead to an increase in openness among academic scientists. In the world of commerce, patents are perceived as promoting openness because they are seen as an alternative to trade secrets, which clearly constitute a more restrictive legal mechanism. A patent offers downstream readers an opportunity to extend a technology by providing details about how an invention works—details that would be unavailable under a regime of trade secrecy. However, in the world of academic science, the restrictions on openness motivated by possible commercial exploitation might tend to propagate upstream from the point of potential patent back into the research process. Consequently, one would expect access to portions of data-streams that are believed to be precursors of potentially patentable products to be relatively tightly controlled.

Second, there is also considerable empirical evidence that intellectual property considerations actually reduce openness, at least on occasion. Michael Mackenzie, Peter Keating, and Alberto Cambrosio, for example, show how the expansion of what was considered patentable in the realm of hybridoma and monoclonal antibodies was accompanied by reductions in the free flow of scientific information (Mackenzie and others 1990). My own interviews with molecular geneticists suggest that at least minor delays in publication sometimes occur while scientists, university technology-transfer offices, and patent lawyers make assessments of the potential commercial value and patentability of results. Survey research on academic-industrial relations has also suggested that biotechnology faculty with commercial involvements are more likely to have engaged in practices that restrict scientific openness (Blumenthal 1992) and some evidence suggests that concerns about intellectual property protection can complicate negotiations about scientific exchange and in some cases make it more difficult to form and maintain collaboration. Indeed, one way to interpret efforts to develop and put into use standard "material-transfer agreements" is as an attempt to reduce the impact of such complexities on scientific collaboration.

A third point about the likely effects of intellectual property considerations on academic science is that the effects will not be uniform across all scientific fields. That is true not only in the trivial case of comparing lines of research with different commercial potential, but much more generally. The perspective outlined above suggests that the effects of intellectual property concerns will be mediated by the prevailing structure of data-streams in particular lines of research. Access practices are probably most intensively shaped not at the level of the discipline or field, but at the level of much-narrower links of research, such as disease-gene mapping, that can be defined in terms of a characteristic data-stream and a particular competitive structure.

If that is true, then one might ask in which lines of research one would expect to find intellectual property considerations producing the largest reductions in openness. The data-stream perspective suggests that the answer might depend in large part on the specific competitive structure of a field of research. In a field characterized by races with intense zero-sum competition, commercial concerns will probably not have a pronounced effect; even in the absence of the potential for profit, the reasons for restricting access are already strong. For example, even if disease genes could not be patented, the winner of the race to find an important gene in the late 1980s could expect substantial rewards. At that time, few human disease genes had been cloned, and cloning one constituted a major achievement.

However, it is important to recognize that intense zero-sum competition is not the typical situation in academic research. In many fields, scientific goals might not be sufficiently well defined, agreed on, and focused on identifiable targets to inspire races among rivals with the same finish line in mind. Instead, research might be exploratory; and in some cases, only one laboratory might be pursuing a given line of investigation. In the absence of focused competition among research groups, openness might be relatively unrestricted. Consequently, one might expect the greatest reductions in academic openness to be provoked by introducing the prospect of commercialization into less-competitive situations.

REFERENCES

Blumenthal D. 1992. Academic-industry relationships in the life sciences: extent, consequences, and management. J Amer Med Assoc 268(23): 3344-3349.

Callon M. 1995. Four models for the dynamics of science. In: Jasanoff S, et al., editors. Handbook of science & technology studies. Newbury Park, CA: Sage Publications. Pp 29-63.

Collins HM. 1985. Changing Order: Replication and Induction in Scientific Practice. University of Chicago Press.

Collins HM, and Pinch, T. 1993. The Golem. Cambridge Univ Press.

Fujimura JH. 1987. Constructing 'do-able' problems in cancer research. Social Studies of Science 17: 257-293.

Hagstrom WO. 1965. The scientific community. New York: Basic Books.

Hilgartner S. (in press). Data access policy in genome research. In: Thackray A, editor. Private science. University of Pennsylvania Press.

Hilgartner S and Brandt-Rauf SI. 1994. Data access, ownership, and control: toward empirical studies of access practices. Knowledge: Creation, Diffusion, Utilization 15(4): 355-72.

Hilgartner S. 1995. Biomolecular databases: new communication regimes for biology?" Sci Comm 17(2): 240-63.

Hilgartner S, Brandt-Rauf SI. in press. Controlling data and resources: access strategies in molecular genetics In: David P and Steinmueller E, editors. A productive tension: university-industry research collaborations. Stanford University Press.

Knorr-Cetina KD. 1981. The manufacture of knowledge. New York: Pergamon.
Latour B. 1987. Science in action. Cambridge, MA: Harvard University Press.
Latour B and Woolgar, S. 1979. Laboratory life. Beverly Hills, CA: Sage Publications.
Lynch M. 1985. Art and artifact in laboratory science. London: Routledge and Kegan Paul.
Mackenzie M, Keating P, Cambrosio A. 1990. Patents and free scientific information: making monoclonal antibodies proprietary. Sci Tech Human Values 15(1): 6-83.
McCain KW. 1991. Communication, competition, and secrecy: the production and dissemination of research-related information in genetics. Sci Tech Human Values 16(4): 491-516.
Merton RK. 1973 [1942]. Science and technology in a democratic order. reprinted as The normative structure of science. In Merton RK, editor. The sociology of science. Chicago: University of Chicago Press.
Ross M, Sicoly F. 1982. Egocentric biases in availability and attribution. In: Kahneman D, et al., editors. Judgments under uncertainty: heuristics and biases. Cambridge Univ Press.

5

Case Studies

INTRODUCTION

Each of the following cases involves an important research tool in molecular biology, and each was chosen to illustrate a form of protection of intellectual property and a pattern of development involving both the public and the private sector. For each case, we present background material and a summary of the discussion that raised issues peculiar to the case.

The ideal strategies for the handling of intellectual property in molecular biology are not always immediately obvious, as these case studies illustrate. For most, final decisions have not been made about how access to these research tools will be controlled. Such decisions might be modified in response to both scientific and legal developments.

RECOMBINANT DNA:
A Patented Research Tool, Nonexclusively Licensed With Low Fees

The Cohen-Boyer technology for recombinant DNA, often cited as the most-successful patent in university licensing, is actually three patents. One is a process patent for making molecular chimeras and two are product patents—one for proteins produced using recombinant prokaryote DNA and another for proteins from recombinant eukaryote DNA. Recombinant DNA, arguably the defining technique of modern molecular biology, is the founding technology of the biotechnology industry (Beardsley 1994). In 1976, Genentech became the first company to be based on this new technology and the first of the wave of biotechnology companies, which in fifteen years has grown from one to over 2000.

The first patent application was filed by Stanford University in November 1974 in the midst of much soul-searching on the part of the scientific community. Stanley Cohen and Herbert Boyer, who developed the technique together at Stanford and the University of California, San Francisco (UCSF), respectively, were initially hesitant to file the patent (Beardsley 1994). Several years of discussion involving the National Institutes of Health (NIH) and Congress followed. By 1978, NIH decided to support the patenting of recombinant DNA inventions by universities; in December 1980, the process patent for making molecular chimeras was issued. The product patent for prokaryotic DNA was issued in 1984. The patents were jointly awarded to Stanford and UCSF and shared with Herbert Boyer and Stanley Cohen. The first licensee signed agreements with Stanford on December 15, 1981. As of February 13, 1995, licensing agreements had generated $139 million in royalties, which have shown an exponential increase in value since their beginning. In 1990-1995 alone, the licensing fees earned $102 million.

This case has three key elements. First, the technology was inexpensive and easy to use; from a purely technical standpoint, there were only minimal impediments to widespread dissemination. Second, there were no alternative technologies. Third, the technology was critical and of broad importance to research in molecular biology.

The technology was developed in universities through publicly funded research. The strategy used to protect the value of the intellectual property was to make licenses inexpensive and attach minimal riders. The tremendous volume of sales made the patent very lucrative. Every molecular biologist uses this technology. However, not all inventions are as universally critical. Only a few university patents in the life sciences, such as warfarin and Vitamin D, have been even nearly as profitable as the Cohen-Boyer patent. Clearly, had this technology not been so pivotal for molecular biology or had an equally useful technology been available, the licenses would not have been sold so widely and the decision to license the technology might have met with more resistance.

The Cohen-Boyer patent is considered by many to be the classic model of technology transfer envisaged by supporters of the Bayh-Dole Act, which was intended to stimulate transfer of university-developed technology into the commercial sector. Ironically, it presents a different model of technology than that presumed by advocates of the Bayh-Dole act (for discussion, see chapter 3). Lita Nelsen, director of the Technology Licensing Office at the Massachusetts Institute of Technology (MIT), noted that the premise of the Bayh-Dole Act is that exclusivity is used to induce development and that universities should protect their intellectual property because without that protection, if everybody owns it, nobody invests in it. "The most-successful patent in university licensing, in the entire history of university licensing, is the Cohen-Boyer pattern which is just the reverse. It is a nonexclusive license. It provides no incentive, just a small tax in the form of royalties on the exploitation of the technology."

The biotechnology boom that followed the widespread dissemination of recombinant DNA techniques transformed the way universities manage intellectual property. It also fundamentally changed the financial environment and culture of biological research.

Nelsen described two ways in which this patent was so successful in fostering the aims of the Bayh-Dole Act. First, it got the attention of biologists by showing the advantages of protecting intellectual property. Stanford earned respectability for the venture by involving NIH and discussing in a public forum how this technology could be disseminated in a way that would not impede research. Second, it got the attention of university chancellors. They began to see that licensing, patenting, and technology transfer might have some financial benefits for the university. Nelsen commented that "that went a little too far. Everybody was waiting for $100 million per year out of their technology transfer offices. Most of them did not get it, and most of them are never going to get it." In the meantime, technology transfer managers developed more experience and became professionalized. They began to learn how to decide what to patent, how to market technology, and how to close deals at reasonable prices and with reasonable expectations. And industry learned how to negotiate licenses with universities.

Nelsen concluded that the whole biotechnology industry came out of the Cohen-Boyer patent, not only because Cohen-Boyer developed gene splicing, but because universities learned how to do biotechnology and early technology licensing—even if the first example was paradoxical.

The decision to negotiate nonexclusive, rather than exclusive, licenses was critical to the industry. If the technology had been licensed exclusively to one company and the entire recombinant DNA industry had been controlled by one company, the industry might never have developed. Alternatively, major pharmaceutical firms might have been motivated to commit their resources to challenging the validity of the patent.

Nelsen noted that at most major universities, it has become standard in industry-sponsored research agreements that the university will retain ownership of any resulting patents but almost without exception will grant the sponsor a first option to an exclusive license. With the increase in university-industry partnerships this applies to more research than in past years. Moreover, the Bayh-Dole Act encourages universities to grant exclusive licenses to companies even if the research was publically sponsored. But as the next case study shows, even when a company holds exclusive rights to a fundamental technology, it might choose to disseminate the technology broadly.

PCR AND *TAQ* POLYMERASE:
A Patented Research Tool for Which Licensing Arrangements Were Controversial

Polymerase chain reaction (PCR) technology presents an interesting counterpoint to the Cohen-Boyer technology. Both are widely used innovations seen by many as critical for research in molecular biology. However, the licensing strategies for the two technologies have been quite different, and they were developed in different contexts.

PCR allows the specific and rapid amplification of targeted DNA or RNA sequences. *Taq* polymerase is the heat-stable DNA polymerase enzyme used in the amplification. PCR technology has had a profound impact on basic research not only because it makes many research tasks more efficient, in time and direct cost, but also because it has made feasible some experimental approaches that were not possible before the development of PCR. PCR allows the previously impossible analysis of genes in biological samples, such as assays of gene expression in individual cells, in specimens from ancient organisms, or in minute quantities of blood in forensic analysis.

In less than a decade, PCR has become a standard technique in almost every molecular biology laboratory, and its versatility as a research tool continues to expand. In 1989, *Science* chose *Taq* polymerase for its first "Molecule of The Year" award. Kary Mullis was the primary inventor of PCR, which he did when he worked at the Cetus Corporation. He won a Nobel Prize for his contributions merely 8 years after the first paper was published in 1985, which attests to its immediate and widely recognized impact. Tom Caskey, senior vice-president for research at Merck Research Laboratories and past-president of the Human Genome Organization, attributes much of the success of the Human Genome Project to PCR: "The fact is that, if we did not have free access to PCR as a research tool, the genome project really would be undoable. . . Rather than bragging about being ahead, we would be apologizing about being behind."

Whereas recombinant DNA technology resulted from a collaboration between university researchers whose immediate goal was to insert foreign genes into bacteria to study basic processes of gene replication, PCR was invented in a corporate environment with a specific application in mind—to improve diagnostics for human genetics. No one anticipated that it would so quickly become such a critical tool with such broad utility for basic research.

Molecular biology underwent considerable change during the decade between the development of recombinant DNA and PCR technologies (Blumenthal and others, 1986). The biotechnology industry emerged, laws governing intellectual property changed, there was a substantial increase in university-industry-government alliances, and university patenting in the life sciences increased tenfold (Blumenthal and others 1986, Henderson and others). There was virtually no controversy over whether such an important research tool should be patented and no quarrel with the principle of charging licensing fees to researchers. The controversy has been primarily over the amount of the royalty fees.

Cetus Corporation sold the PCR patent to Hoffman-LaRoche for $300 million in 1991. In setting the licensing terms for research use of PCR, Roche found itself in a very different position from Stanford with respect to the Cohen-Boyer patent. First, it was a business, selling products for use in the technology. That made it possible to provide rights to use the technology with the purchase of the products, rather than under direct license agreements, such as Stanford's. This product-license policy was instituted by Cetus, the original owner of the PCR patents. An initial proposal to the scientific community by the president of Cetus for reach-through royalties—royalties on second-generation products derived through use of PCR—was met with strong criticism. Ellen Daniell, director of licensing at Roche Molecular Systems, noted that the dismay caused by the proposal has continued to influence the scientific community's impression of Roche's policy.

Roche's licensing fees have met with cries of foul play from some scientists who claim that public welfare is jeopardized by Roche's goals. Nevertheless, most scientists recognize that Roche has the right to make business decisions about licensing its patents. The fact that Roche had paid Cetus $300 million for the portfolio of PCR patents led some observers to think that Roche intended to recoup its investment through licensing revenues, a point that Daniell disputed. She pointed out that Roche's business is the sale of products and that licensing revenues are far less than what would be needed to recoup the $300 million over a time period that would be relevant from a business viewpoint. Daniell listed Roche's three primary objectives in licensing technology:

- Expand and encourage the use of the technology.
- Derive financial return from use of the technology by others.
- Preserve the value of the intellectual property and the patents that were issued on it.

Roche has established different categories of licenses related to PCR, depending on the application and the users. They include research applications, such as the Human Genome Project, the discovery of new genes, and studies of gene expression; diagnostic applications, such as human in vitro diagnostics and the detection of disease-linked mutations; the production of large quantities of DNA; and the most extensive PCR licensing program, human diagnostic testing services. Licenses in the last-named category are very broad; there are no up-front fees or annual minimum royalties, and the licensees have options to obtain reagents outside Roche.

Discussion about access to PCR technology centered on the costs of *Taq* polymerase, rather than on the distribution of intellectual property rights. Tom Caskey's view was that "the company has behaved fantastically" with regard to allowing access to PCR technology for research purposes. Bernard Poiesz, professor of medicine at the State University of New York in Syracuse and director of the Central New York Regional Oncology Center, agreed that he knew of no other company that had done as well as Roche in making material available for research purposes. But he also argued that the price of *Taq* polymerase is too high and has slowed the progress of PCR products from the research laboratory to the marketplace. Poiesz stated that the diagnostic service licenses "are some of the highest royalty rates I have personally experienced." He cited the example of highly sensitive diagnostic tests for HIV RNA, which he said are too expensive for widespread use, largely because of the licensing fees charged by Roche.[1] Caskey felt that Roche should have expanded the market by licensing more companies to sell PCR-based diagnostic products and profited from the expansion of the market, rather than from the semiexclusivity that it has maintained.

Nor are all university researchers satisfied with their access to *Taq* polymerase. Ron Sederoff commented that—in contrast to the human genomics field, in which funding levels are much higher than for other fields of molecular biology—many academic researchers do not find easy access to the technology. Several workshop participants noted that the high cost *Taq* polymerase made many experiments impossible for them.

What is the effect of the Cetus-Roche licensing policy on small companies? Tom Gallegos, intellectual property counsel for OncoPharm, a small biotechnology company, stated that most small companies cannot afford the fees charged by Roche. He noted that the entry fee for a company that wants to sell PCR-based products for certain fields other than diagnostics ranges from $100,000 to $500,000, with a royalty rate of 15%. By comparison, a company pays about $10,000 per year and a royalty fee of 0.5-10% for the Cohen-Boyer license. The effect is an inhibition of the development of PCR-related research tools, with consequent reductions or delays in the total royalty stream and possibly litigation.

Sidney Winter, professor of economics at the Wharton School of Business, suggested that in asking whether the price of some technology is too expensive, one should consider "compared with what?" Compared with licensing and royalty fees for Cohen-Boyer, PCR might seem excessive. If one imagines that the cost of the PCR patent were financed by a tax on the annual US health-care expenditure which was about $1 trillion in 1995 (Source: Congressional Budget Office), that tax would be roughly equal to 0.03% and might be a price worth paying for the advances made possible by PCR technology.

During the workshop, several people distinguished between research tools

---

[1] Royalty rates refer to a charge based on the revenues earned by the licensee and are different from the up-front fees and annual minimum royalties referred to earlier. As a member of a not-for-profit institution, Poeisz was offered the choice between a 9% or 12% royalty rate, with the lower rate available to those who agreed to use Roche-manufactured DNA polymerase for their testing.

that are commercial products and tools that have little market value but are important tools for discovery. In the case of PCR, the research tool is both a commercial product and a discovery tool. As such, it raises questions. Are the PCR patents an example of valuable property that would have been widely disseminated in the absence of patent rights? Is PCR an example of a technology that has been more fully developed because of the existence of patent rights? Daniell stated that Roche has added considerable value to the technology, in part through the mechanism of patent rights. There was vigorous discussion and disagreement as to whether the licensing fees justify the value added by Roche.

PROTEIN AND DNA SEQUENCING INSTRUMENTS:
Research Tools to Which Strong Patent Protection Promoted Broad Access

This case study was selected because it provides a clear example of how patent protection promoted the development and dissemination of research tools. By most standards, this would be considered a successful transfer of technology. The possibility of automated, highly sensitive DNA and protein sequencers was developed in the public sector by Leroy Hood's group at California Institute of Technology (Cal Tech). However, it was only with the help of substantial private investment that these research tools were widely disseminated.

The ability to synthesize and sequence proteins and DNA revolutionized molecular biology; automating these tasks promised to consolidate the revolution. Indeed much of the achievement of the Human Genome Project is attributable to the development of automated sequencing instruments, which greatly reduced the time and cost needed to sequence DNA. Because the effects of genes depend on the proteins that they encode, protein sequencing has been a key step in deciphering gene function. Until automated sequencing instruments were widely available, only a few laboratories had access to this technology.

The prototypes for these instruments were developed in Hood's laboratory during the years 1970 - 1986. Over a period of six or seven years, the team of scientists assembled by Hood increased the sensitivity of protein sequencing instruments by a factor of about 100. That transformed a difficult and uncertain task into one that could be reliably accomplished with the minute quantities of purified proteins that so often limited the scope of the analysis. Hood's laboratory was the first to sequence lymphokines, platelet-derived growth factor, and interferons. After those successes, he was approached by many scientists who asked why the technology could not be made available to the whole research community. Since the middle 1990s, the technology has become widely available.

The broad availability of sequencing technology is due, in no small part, to Hood's perseverance in the face of widespread skepticism. His 1980 manuscript that described, for the first time, automated DNA sequencing was delayed by the journal *Nature* on the grounds that this technology sounded like "idle speculation." Hood wrote three or four proposals to NIH and the National Science Foundation but was unable to obtain funding for his instrumentation work. The bulk of the support for this technology came from the private sector, and even then companies were reluctant to invest in developing the sequencing instrumentation. He approached nineteen companies, all of which declined to support the development of the sequencers. Eventually, he obtained funding from Applied Biosystems (ABI), but even this support required difficult negotiations between Cal Tech and ABI. ABI insisted on, and received, an exclusive license. As Hood told it, the argument that convinced Cal Tech to support the arrangement was that "if the scientific community wants these instruments, it is our moral obligation to make them commercially available."

At the time of this workshop, ABI had sold more than 3,000 DNA sequencers and more than 1,000 protein sequencers worldwide (although some elements of the technology, such as peptide synthesis, were not protected by patents, most of the instrumentation was patented by ABI). Sequencing facilities that serve multiple investigators are now standard features at research universities. That is not to say that licensing of this technology has been without controversy. Cal Tech licensed the technology to ABI with the stipulation that ABI would sublicense it under what Cal Tech considered reasonable terms. A number of companies have argued that ABI's terms are not reasonable. As with PCR, the situation is complicated in that the primary licensee claims that its license fees reflect what it needs to charge to earn a reasonable return on its investment in developing the technology.

ABI is clearly the leader in the world market for DNA sequencers. But other companies, such as Pharmacia and LI-COR, have important market shares. LI-COR has established a niche in the market with its infrared fluorescence DNA sequencer; infrared light has low background fluorescence, which allows for the development of more robust, solid-state instrumentation than is possible with other DNA sequencing technology. LI-COR is typical of many small biotechnology companies in its reliance on its patent portfolio. Harry Osterman, director of molecular biology at LI-COR, noted that "DNA sequencing is more than just an instrument, it is a system. To make a viable product, all the disparate pieces need to be integrated. That makes for a challenging intellectual property and licensing exercise, unless you have the internal funds to do everything. You require instrumentation, software, chemistry, and microbiology." Patent protection allows a small company to negotiate cross-licenses, which are critical in systems technologies, such as sequencing instrumentation. It can provide an opportunity that a small company would not otherwise have to compete in a market.

One might argue that patent protection served both the large company (ABI) and the small company (LI-COR) in bringing their sequencing technology to the market. In the case of ABI, patent protection afforded them the opportunity to develop a complex system of technology in an orderly and efficient manner, as proposed by the prospect development theory presented by Richard Nelson in chapter 3. In the case of LI-COR, patent protection of sequencing systems enabled it to negotiate the cross-licenses needed to develop its product fully. In both cases, private support has driven the development and dissemination of a research tool. The public and private sectors seem to have gained equally.

RESEARCH TOOLS IN DRUG DISCOVERY:

Intellectual Property Protection For Complex Biological Systems

Research tools in drug discovery present an example of the difficulties in protecting intellectual property when technologies involve complex biological systems that lack discrete borders. The information is often broad and refers to general categories of matter, such as a class of neural receptors, rather than finite entities, such as the human genome, or specific techniques, such as PCR or recombinant DNA techniques. Controversies have emerged over broad patents, which some see as stifling research on and development of useful drugs and others see as critical to the translation of research knowledge into useful products. The focus of the discussion in the workshop was the tension between the dependence of small biotechnology companies on patents and the difficulties created when research on complex biological systems is restricted by a thicket of patents on individual components of the systems.

When research on a complex system—for example, receptor biology or immunology—requires obtaining multiple licenses on individual components of the system, the potential for paying substantial royalty fees on any useful application derived from that product can be daunting. "Royalty stacking" can swamp the development costs of some therapies to the point where development is not economically feasible. That is a problem particularly in gene therapy, where the most promising advances now are related to rare genetic diseases that present small markets.

Bennett Shapiro, vice-president for worldwide basic research at Merck Research Laboratories, argued that the central issue is not about patenting, but about access, about encouraging the progress of biomedical research. Problems can arise when access to related components of biological systems is blocked. For example, schizophrenia is often treated with compounds that suppress dopaminergic neurotransmission. Many such compounds, for example haloperidol, act nonspecificially and suppress the entire family of dopamine receptors. People who take those compounds for schizophrenia often develop other disorders some of which resemble Parkinson's disease, another disease involving the dopamine system. A rational approach to discovery of improved schizophrenia drugs would be to target specific dopamine receptors. But if different companies hold patents on different receptors, the first step on the path to an important and much needed therapeutic advance can be blocked.

Shapiro commented that when only one company starts along the path of discovering a particular type of drug, its chance of discovering it is very low. Merck supports only a tiny fraction of total biomedical research, and it benefits enormously from research going on elsewhere in the world. It is in Merck's interest to share the results of its research with the understanding that they can be even more useful if placed in the pool of worldwide research resources.

It is interesting to compare that perspective on drug discovery with the early history of radio and television, other examples of complex systems of which many components were patented individually. In chapter 3, Richard Nelson noted that it was not until cross-licensing practices became widespread in the early development of radio and television that important advances that enabled broad access to the technology took place. When the intellectual property was sequestered in the hands of a few companies, the entire electronics industry remained sluggish. Of course, the progress of the industry overall must be balanced by the financial needs of individual companies. Shapiro noted that Merck has felt the need to become more energetic about patenting than it was years ago. For example, carrageenan footpad assays were used to develop non-steroidal anti-inflammatory drugs. The assays were in the public domain, and many companies used them to develop new drugs. Today, Merck would patent such an assay and use its patent position to trade with other companies for access to other research tools.

James Wilson, director of the Institute for Human Gene Therapy at the University of Pennsylvania, described his experience with the different ways in which patents on research tools are used. One is to block others from using the tools—to protect one's proprietary use—which he did not see as economical. Genetic therapy patents might not generate enough financial return to offset the investment costs. Wilson also suggested that genetic therapy patent files are only going to waste money in lawsuits brought against those patents. Second is to generate revenues for universities to support their infrastructures, although, as Lita Nelsen noted, most universities are not likely to earn much from patent revenues. The third is to barter so as to continue development without creating an economic disadvantage.

Like previous panelists, Larry Respess of Ligand Pharmaceuticals, argued that the chances of survival of a small biotechnology company would be slight without patents. He noted that the biotechnology industry is composed of small companies that have grown through venture capital and public offerings and that finance research through equity, not product revenues. The goal is to develop products and then evolve into an independent company.

Wilson also pointed to a dramatic increase over the last two to three years in the difficulty in transferring material between universities. Nelsen emphasized that university technology transfer managers are still learning. And many decisions of the US Patent and Trademark Office (PTO) are controversial and under close scrutiny by those charged with managing intellectual property.

In commenting that "it is hard to know what the proprietary landscape is going to be, but it will be complex, whatever it is," Wilson summarized many of the workshop participants' comments.

Changes in Biotechnology Strategies

Respess discussed how R & D strategies for biotechnology have changed over the last twenty years. The biotechnology industry was born in about 1975 by Genentech, and most of the companies that followed Genentech pursued a similar strategy. Their objective was to produce and sell therapeutically-active large protein molecules, which was made possible by the availability of the Cohen-Boyer technology. The strategy was to discover and try to patent a gene for such a protein; it was hoped that the gene could be used to express abundant quantities of the protein. Some of the early examples are insulin, growth hormone, erythropoietin, and the interferons.

The advantages of that approach were that everyone knew that the products would be useful and that recombinant techniques were efficient for production, compared with earlier techniques of extraction from cadavers and tissue. Another advantage —albeit not from a scientific viewpoint—is that it is easy to sell to the investment community; it was a simple, easily understood model. Respess described the raising of capital in the early days of biotechnology as "unbelievable. You could found a company and, within a relatively short time, go public and raise many millions of dollars." However, those days are now past, in part because of the intrinsic limitation of large protein molecules: they are expensive to produce and to deliver to patients (they must be delivered by injection). The drug targets that are easy to identify have already been exploited.

A newer biopharmaceutical strategy emerged—not to discover large proteins or other large-molecule drugs, but to find other therapeutically active small molecules. These are the traditional targets of pharmaceutical research, but a biopharmaceutical company uses modern biotechnology and insights from molecular biology to get to the ultimate target product more quickly and efficiently. This approach has several advantages. The drugs are conventional and can typically be given orally, as well as by injection; they are relatively easy to manufacture; and the Food and Drug Administration is very familiar with such drugs, which makes it easier to get a new drug approved. The problem from a small company's perspective, however, is that it takes a very expensive infrastructure. Ultimately, synthesizing small molecules means making many molecules, and medicinal chemistry is very expensive. You have a tool, but you do not have any products in hand.

EXPRESSED-SEQUENCE TAGS (ESTs):
Three Models for Disseminating Unpatented Research Tools

An expressed-sequence tag (EST) is part of a sequence from a cDNA clone that corresponds to an mRNA (Adams and others 1991). It can be used to identify an expressed gene and as a sequence-tagged site marker to locate that gene on a physical map of the genome. In 1991 and 1992, NIH filed patent applications for 6,800 ESTs and for the rapid sequencing method developed by Craig Venter, who was a scientist at NIH. The PTO rejected NIH's application and when Harold Varmus became director of NIH, he decided not to appeal. But controversy caused by the initial patent application continued. In 1992, Venter left NIH to form The Institute for Genome Research (TIGR), a nonprofit company, and William Haseltine joined the newly established private company, Human Genome Sciences (HGS), a for-profit company that initially provided almost all of TIGR's funding. The focus of the controversy then moved from the public to the private sector, and it changed from an issue about patenting research tools to an issue of access to unpatented research tools. Like many other research tools, ESTs fill different roles and some of the controversy has involved disputes of the relative importance of ESTs for uses other than research.

Two factors have contributed to the controversy over intellectual property issues in this particular setting. First is the perception that some of the participants have been staking out intellectual property claims that extend beyond their actual achievements to include discoveries yet to be made by others. There is no question that ESTs constitute a powerful research tool. Questions about the patenting of ESTs have focused on the criteria of utility. ESTs are of limited value without substantial and nonobvious development. Initially a public institution, NIH, proposed to patent discoveries that both scientists and some representatives of industry felt belonged in the public domain. More recently a private institution, Merck, has assumed the quasigovernment task of sponsoring a university-based effort to place information into the public domain. While other private companies have provided funds for public sector research, such as in the Sandoz-Scripps agreement, these efforts have not been with the expressed purpose of putting information into the public domain.

This is a particularly interesting case study, in part because it began as a controversy over patents—over what could be patented, what should be patented and what would be the effect of patenting. It has evolved into a controversy over the dissemination of unpatented information and the terms on which that information will be made available.

Different firms have taken different approaches to the dissemination of these unpatented research tools, thus providing a natural experiment with which to study three models for disseminating the same sort of information. The models all arose in the private sector, and we can assume that although each firm adopted a different strategy, they had the same ultimate goal of maximizing the value that they could obtain from the information. Merck has put the information in the public domain, Human Genome Sciences (HGS) initially adopted an exclusive-licensing model, and Incyte adopted a broad licensing approach of offering non-exclusive licenses to its database to as many firms as would sign up. Putting information in the public domain limits opportunities to exploit it as a trade secret by controlling access to it. Patents, or the patent applications of private database owners, potentially limit the ability to use the information that is in the public domain if any patent rights are ultimately obtained.

- *HGS.* The strategy of HGS has been to form a major partnership with the pharmaceutical firm SmithKline Beecham (SKB) [2], with which it agreed to provide a three year exclusive license to its EST database. SKB has sublicensed its rights to a major Japanese pharmaceutical company, Takeda Chemical, Ltd., and HGS also has 200 restricted-licensing arrangements with university researchers. The TIGR database contains a limited portion of the data created by HGS and all of the data created by TIGR before April 1, 1994 which is when TIGR stopped work on human cDNAs. TIGR provides two levels of access to its EST databases. At the first level, an investigator is allowed access to sequences that are owned by HGS that overlap or are identical with sequences already in the public domain and for which public databases are available. At the second level, investigators are allowed access to about 70,000 sequences that are not listed in the publicly available databases (GenBank or the European databases). To obtain the second level, an investigator must agree to disclose any invention that is made at any time after access is gained. Furthermore, HGS or the Institute for Genome Research (TIGR) must be allowed at least six months to negotiate a licensing agreement.[3] The public does not have access to the much larger HGS cDNA database.
- *Merck.* Merck is interested in using the information from ESTs for furthering its research efforts. The Merck Gene Index was established to fill a public-access gap and was developed in partnership with established genome centers. Sequencing is carried out at Washington University, and the data are handled at the Los Alamos Laboratories. The international databases are a direct source of the information. A biotechnology company has taken all the clones into its distribution system and will freely distribute its materials. Other institutions, such as TIGR and Genethon, have entered sequences into this public database.

---

[2] Takeda Chemical, Ltd., the largest Japanese pharmaceutical firm, is another partner. Since the workshop, HGS has directly licensed three other companies: Schering-Plough, Merck KGAA (a German company, not affiliated with US Merck), and Synthelabo (a French company).

[3] After April 1, 1997, all of the original EST sequences in the HGS-TIGR databases completed by April 1994 will be publicly available with no restrictions.

- *Incyte*. Incyte's strategy has been to offer nonexclusive licenses to its database. As of the time of the workshop, six companies (Pfizer, Upjohn, Novo Nordisk, Hoechst, Abbott Laboratories, and Johnson & Johnson) have contributed in the aggregate, around $100 million, exclusive of contingency payments and royalty payments for access to this database. Even as the Merck data continue to be placed in the public domain, Incyte continues to sign up new subscribers; there seems to be continuing value for the subscribing firms to obtain access to one of the private databases. This strategy is interesting not only for what it says about the nonexclusive-licensing strategy but because this is the most current information as to the relative values of the private databases versus the public-domain database.

The Informational Value of ESTs Is Rudimentary

None of the participants disputed the value of ESTs as research information, but several commented on the rudimentary nature of the information. Having an EST in hand does not guarantee a practical strategy for obtaining the identity of the gene of which the EST is but a fragment. Furthermore, if the gene identified is unknown, there remains substantial investment in understanding its function. It has been successfully accomplished in many cases, and many specific strategies have been developed over the years for approaching this task. Nonetheless, it remains fraught with uncertainty. In 1995 the Human Genome Organization (HUGO) issued a statement on "Patenting of DNA sequences" arguing that the nature of sequence information is so rudimentary that to limit access to it is to impede development of medical advances.

Several uses have been suggested for genes and gene fragments to claim utility requirement for patent protections. They include the use of genes or gene fragments for categorizing, mapping, tissue typing, forensic identification, antibody production, or locating gene regions associated with genetic disease. However, each of those suggested uses may not be carried out without considerable further effort and additional biological information that is not inherent in the sequence alone. Many of the workshop participants concurred with the HUGO statement that without databases to provide further information, the informational value of ESTs themselves is very limited.

William Haseltine, CEO of HGS, noted that patent applications filed by HGS for ESTs involve considerably more than simply identification of the gene fragments and involve information about the stage of development and tissue type in which those genes are expressed. He further commented that the importance of the EST database is not simply that the fragments are identified, but that the database itself provides a high level of information.

The Value of ESTs Could Be Reduced by Limiting Access

Many of the workshop participants echoed the HUGO statement of concern that "the patenting of partial and uncharacterized cDNA sequences will reward those who make routine discoveries but penalize those who determine biological function or application. Such an outcome would impede the development of diagnostics and therapeutics." Both Harold Varmus and Gerald Rubin suggested that some researchers are likely to be discouraged from working on patented ESTs for fear that the patent holders would lay claim to their future discoveries, particularly discoveries about gene function, which are clearly of far greater biological utility than the identification of anonymous fragments and are more likely to have useful applications for human health.

Several previous reports have stated that research-tool claims should not be so broad as to block the discoveries outside of the patent (House of Commons Science and Technology Committee 1995, National Academies Policy Advisory Group 1995). No one at the workshop argued otherwise.

Fragile X syndrome, which is the most-common form of mental retardation, provides an example of how ESTs can contribute to human disease. The name refers to the fact that the X chromosome is easily broken. Caskey described how he, Steve Warren, and Ed Benustra used an EST to discover that the genetic defect involves multiple repeats of the nucleotide triple CGG. They went on to characterize the gene, and that provided the information necessary to develop what is now the most widely used diagnostic test for fragile X syndrome. When they made their discovery, the sequence information on the gene involved gave no information on function. It was investigators like Bob Nussbaum, and Dreyfus, at Philadelphia, who went on to identify the gene's function.

Caskey suggested that if speculative claims were permitted among a certain set of ESTs the rights of investigation to discover that gene would be denied.

James Sikella cited the example of the HIV patent, which is jointly held by the US and French governments. The patent has not been tightly restricted for investigational use. At the time of its filing, its sequence and functions were not described. Many discoveries about HIV have evolved from that sequence information, and Sikella noted that it would have been a disservice to the public if the sequence information had not been available as a general research tool.

The Human Genome Is Finite

As of this workshop, some 27,000-35,000 human genes were represented in the database. Humans are estimated to have about 80,000-100,000 genes, so that represents about one-fourth to almost half of the total. Tom Caskey predicted that as the database begins to be flooded with sequence information, there will be a higher stringency on patents and patent claims will be directed more toward functional aspects of the genes, rather than being primarily descriptive.

Caskey also described how the usefulness of the gene index has improved with the addition of more sequences. When the general location of a disease gene is known from genetic mapping, limited sequencing is an important strategy for finding the gene. By sampling the critical region, the small bits of sequences can be used to search for homologies in the gene sequence database. In this way, a previously sequenced gene or gene fragment can be identified as being located in a critical region. Such a gene is then a prime suspect for more detailed studies in those individuals carrying the disease. Initially, the success rate for this technique of finding disease genes in positionally cloned regions was only about 40%. As the size of the gene database increases, so does the success rate. This is, therefore, becoming a fast and facile method for identifying a disease gene in a critical region identified by genetic mapping.

Sikella suggested that the success of the Human Genome Project may be measured, in part, by how the knowledge that it generates benefits society. He emphasized the importance of making these benefits available in a cost-effective way.

The Advent of DNA Sequencing Presents Important Questions About Patentability

Leon Rosenberg commented that "although the debate seems to have cooled a bit, the issues surely have not been resolved." Tom Caskey of Merck and William Haseltine of HGS both commented that they have no quarrel with the current criteria for patents, but they express different views as how those criteria should be interpreted. Since the workshop, HGS has received patents on a number of ESTs with broader claims of utility than the initial EST patent applications filed by NIH in 1974. Whether this will influence the debate over ESTs is an open question. Caskey noted that after one has an EST, identifying the full length sequence cDNA is the obvious next step. And yet this rarely leads to precise knowledge of that gene's function. He predicted that the complete cDNA sequences might become the 1997 version of ESTs—that is, research tools which many people do not believe meets the full potential criteria of novelty, nonobviousness, and utility. Rosenberg suggested that "the biomedical research community has not yet truly grappled with the possibility that a large number of genes could be controlled by the rights of a relatively small number of parties who could not possibly hope to fully exploit their potential value." He suggested that if research tools are not made available to the scientific community and others, we will have to confront this issue directly, whether that requires changes in patent law or other equally drastic directions.

REFERENCES

Adams MD, Kelley JM, Gocayne JD, Dubnick M, Polymeropoulos MH, Xiao H, Merril CR, Wu A, Olde B, Moreno RF, Kerlavage AR, McCombie WR, Venter JC. 1991. Complementary DNA sequencing: expressed sequence tags and human genome project. Science 252(5013): 1651-1656.

Beardsley T. 1994. Big time biology. Sci Amer. November: 90-97.

House of Commons Science and Technology Committee. 1995. Human genetics: the science and its consequences, Vol.1. London, UK: House of Commons

National Academies Policy Advisory Group. 1995. Intellectual property and the academic community. London: The Royal Society. 65p.

6

Perspectives From Different Sectors

INTRODUCTION

The protection of intellectual property has been one of the most challenging issues in the recent proliferation of university-industry-government partnerships—largely because costs and benefits associated with protection of intellectual property are distributed unevenly among different sectors. Even though different sectors might share the general goal of providing useful innovations to society, there are vast differences in how people can contribute to this goal. Optimal strategies clearly depend on immediate goals, needs, and opportunities. Within the university, research scientists and technology-transfer managers have different missions. A university research scientist might have goals and needs different from those of a scientist working in industry. Within industry, the best strategies for protecting intellectual property are different for small biotechnology companies and major pharmaceutical companies.

This chapter summarizes a session in which representatives of the three sectors (university, industry, and government) were asked to discuss their concerns about intellectual property rights and research tools from their own perspectives. The session provided a forum to address the issues more thematically than was possible during the case study discussions. Gerald Rubin and Lita Nelsen spoke from their experiences in the university as a research scientist and technology transfer manager, respectively; Leon Rosenberg and Thomas D'Alonzo spoke from their experiences with a major pharmaceutical company and small biotechnology company, respectively; and Harold Varmus discussed his concerns about the protection of intellectual property in government-sponsored research.

UNIVERSITY RESEARCH

Gerald Rubin, University of California, Berkeley

Intellectual property rights and research tools present two main concerns for academic researchers. First is the exchange of materials and ideas in a timely and unencumbered way. Second is the allocation of credit and reward for discoveries. Although intellectual property rights have been a source of great concern, intellectual property law as practiced has generally not encumbered the timely exchange of materials and ideas among university researchers.

In 1917, Thomas Morgan, the progenitor of *Drosophila* genetics research, wrote in strong terms that scientific material and knowledge should be freely and widely circulated. Even into the 1990s, the *Drosophila* genetics research community has continued to support the free exchange of materials. I offer my own experience as an example of contrasts. My collaborators and I recently published a gene that we had isolated using genetic screening in *Drosophila.* We also cloned the mouse and human genes, and these were described in the same article. I received six requests from *Drosophila* workers for these strains, each of which said something like, "I enjoyed your paper. Please send me . . ."— followed by a long list of reagents and then, "Thank you," and that was it. The dozen requests for the mammalian reagents went something like, "We will, of course, consider this a collaboration. Here is exactly what I am going to do with it. . . " followed by all sorts of stipulations. [This exemplifies Stephen Hilgartner's observations in chapter 4 that practices of sharing material information can differ widely—even between closely related subfields. Hilgartner also notes that access practices are probably most intensively shaped not at the level of a discipline, but at the level of relatively narrow subjects of research, where the force of individual personalities has substantial influence on how research is practiced. This may also illustrate another one of Hilgartner's points which is that narrowly focused competition tends to influence practices of sharing. Human and mouse genes are more likely to have commercial value (whether as strategic intellectual property or for their value as products on the market) and might thus increase the level of competition for sharing.]

Some of the problems in the scientific community with the exchange of material are due more to commercialization than to intellectual property laws. There are problems when academic researchers get money from industry and when the industrial collaborators impose constraints on when and where people can distribute materials, but these problems are not with patent law.

Most academic scientists would agree the patent laws can cause problems with respect to the fair allocation of credit and reward. Academic researchers are taught that giving credit where credit is due is a basic tenet of science. Scientists understand that their discoveries are based on the work of many other people and that it is very important to give proper credit in publications and spoken presentations for the work that went beforehand. That is how intellectual royalties are paid. However, in patent law, all the rewards go to the person who gets to the step of establishing utility first; the basic work that led to that point is not compensated. That is just a fact of life. It makes researchers uncomfortable, and it could get worse. For example, if someone actually got to patent ESTs, and I took a gene with a patented EST and discovered its function, I might not get any financial reward for my effort; the person who patented the ESTs would get the reward. That would clearly be a disincentive for most scientists, and we need to worry about it. Short of changing or rewriting the whole patent law, I am not sure what can be done. That seems to me a much more serious concern than problems with the exchange of material.

UNIVERSITY ADMINISTRATION

Lita Nelsen, Massachusetts Institute of Technology

From the perspective of university administrators, the primary reason to license and protect intellectual property is to induce development and thereby make the products of university research available to taxpayers. Nowadays, expectations of economic development resulting from taxpayers' fundamental research spending are much greater. The university also wants to induce development because the business community wants it. When MIT helps to establish small incubator companies, the city of Cambridge gets more jobs and local real-estate values go up.

If the university wants to attract industrial sponsorship of research, it must have an intellectual property program. The faculty also appreciate the benefits of an intellectual property program. They earn royalties. Some of them get a great deal of psychic reward out of seeing their research actually cure disease or become commercially available products. More mundanely, if we have patents that we license to companies that will hire consultants, our faculty have more consulting opportunities. We think it is good for our teaching mission to have our laboratories engaged in real world problems and bringing industry in. Many of our graduates go to work for our licensees.

Many people believe that the role of technology transfer offices is to make money. University-based technology transfer is not a good way to make money. The MIT technology transfer office is seen as one of the more successful and lucrative university licensing offices. But its revenues are relatively modest.

MIT's budget is close to $1 billion a year, which includes $350 million in research at MIT and $350 million in research at Lincoln Laboratories. The gross income of the technology transfer office is $8 million, including all patent expenses and royalties. The net income, which is distributed to inventors and departments and to the institute's general fund, is only about $3 million a year. Maximizing revenues is not our primary goal, although we have to make enough money to survive. Finally, if we are going to be involved in patenting and licensing, we have to make enough money to provide incentives to the faculty to write patent applications. No university president is going to let people do it at a loss when he or she is looking for functions in the university that can be eliminated to save operating costs.

In theory, there are two kinds of research tools, and they are often confused. One is the kind that becomes useful only if it is commercialized and is otherwise only marginally useful—chromatography or PCR, for example. You can make the invention, publish it and let no one have patent rights in it, and people will still be working with little pieces of paper, test tubes, and candles. Or you can give a license, or rights to the intellectual property to a Hewlett-Packard or a Perkin Elmer, for example, which will take the hand-crafted piece of equipment and turn it into a machine that allows people to do something a thousand times faster than they would otherwise have done.

The other kind is what I call a discovery tool. Rather than a tool for doing research, it is something to work on—for example, a gene or a receptor. A discovery tool does not usually need development before researchers can use it as a research tool. A priori, I would argue against patenting them; if they are not to be patented, they should be licensed nonexclusively because the more people using a discovery tool, the better. Ligand Pharmaceuticals provides a counter-example even to that assumption. Although a few people might have worked on receptor assays, haphazardly, Ligand was able to make a major investment in their receptor research with the commitment of hundreds of millions of dollars because of patenting and exclusive licensing.

There are no easy answers to the question of what should and should not be patented. Tradeoffs are always made. If an exclusive license to a research tool is given to an unsuccessful company that is unable to raise money and then stagnates, but does not quite go out of business, the patent rights will block access to the research tool, and that would be a scandal. If an exclusive license for an important pharmaceutical receptor is given exclusively to a large pharmaceutical company that proclaims that it wants to disseminate knowledge widely, but whose licensing practices are so difficult that it takes three years to get a sublicense, it is a less-visible scandal, but it is still a scandal. The right way is not clear. As a university administrator, I proclaim that universities are trying to figure out what the right thing to do is and trying to evolve norms based on the balancing of equities. But we do not know how to do it yet.

MAJOR PHARMACEUTICAL COMPANY

Leon Rosenberg, Bristol-Myers Squibb

Attitudes Have Greatly Changed

As late as the early 1980s, there remained a deep suspicion of intellectual property among investigators in academe, other research institutions, and government. Even though they understood that patents rewarded inventors for disclosing the nature of inventions to the public, patent law was not considered a mechanism for promoting dissemination of knowledge. Consequently, many scientists strongly opposed any role that their colleagues might have as founders of or collaborators in commercial enterprises. Some even objected to researchers having the status of inventors on university-owned or government-owned patents. In the opinion of many scientists, such relationships fostered unavoidable and unmanageable conflicts of interest. Today, there is greater appreciation and acceptance of the delicate balance that has now been struck in the laws governing intellectual property protection.

Commerce and Science: Legitimate, yet Competing, Interests

We have witnessed the development of an ever-closer alliance between universities, research institutions, government agencies, and private industry in the field of medical research. These increasingly productive, yet always complex, relationships are an accepted reality in today's medical-research enterprise. Their frequency and complexity, however, continue to highlight the inherent tension between two legitimate, yet competing, interests—the commercial incentive to protect intellectual property and the tradition of open communication and free flow of information within the scientific community. Undoubtedly, the many collaborative efforts that exist today will not subside, and there will be more tomorrow. Funding pressures on all parties will drive them closer together as scientists seek scarce research funds from all potential sources and companies seek to maximize their use of all sources of innovation. Thus, it is incumbent on the academic and government research communities and on private industry to communicate and understand each other's positions better. The necessarily different interests and different cultures on both sides of the equation will continue to present us with complex questions that defy simple answers.

Dissemination of research results is one of the most visible ways in which the competing interests intersect. Once new knowledge is created, researchers want to exchange it freely in the scientific community for replication, evaluation, and use. Publication, whether immediate or delayed, is not always welcomed by parties holding a commercial interest in the research. Nevertheless, most corporate entities recognize that it is critical to attract and retain the cooperation of top scientists. As indicated by Blumenthal and others (1996), it is common for companies to "require academic researchers to keep information confidential to allow a filing of a patent application. Such a requirement is standard practice at most academic institutions." Most companies currently request no more than the 90-day waiting period to make the necessary patent filings. As Blumenthal and others also note, "the current policy of the National Institutes of Health provides that 30 to 60 days is a reasonable period to delay the release of information while such an application is being filed. Such delays are a natural consequence of patent law, which requires that confidentiality be maintained until a patent application is filed." In individual cases, companies may request an extension of time to address unique issues. In my experience, such requests generally are few and are made only after a careful weighing of the scientific benefits and the competitive risks associated with disclosure. In general, the individual researcher or the academic institution with which a company is collaborating can deny the request and thus decline to collaborate if granting the extension of time would adversely affect important research objectives. Certainly, there have been instances of abuse of this general idea. But my experience does not suggest that the abuses are pervasive or even widespread.

My own company, Bristol-Myers Squibb, has a variety of such collaborative research agreements and has seldom required more than 60 days of advance notice. As many companies do, Bristol-Myers Squibb takes reasonable steps to limit the number of requests for extension of the 60-day advance notice. Of course, if we cannot convince investigators in their own institutions of the need for the longer duration of confidentiality, they do not need to concede it.

A pivotal issue on the horizon will be the distribution—under a material-transfer agreement,[1] license, or otherwise—of basic research tools that are not otherwise available to other investigators. Numerous companies, large and small, and universities are struggling with this issue. Among the important considerations is that one person's tool can be another person's product. For example, a gene can be a product in the hands of the gene therapy company or a manufacturing system for a company that sells the gene product. The same gene could be a research target for yet another company seeking a small-molecule drug. If the gene was secured only after years of investigation, a commercial enterprise generally would be unwilling to distribute it simply for the asking. Directors of corporate research charged with the responsibility for the investment of their company's shareholder dollars would have to think very hard before recommending a course of action that could result in widespread availability of that target in the community at large.

[1]Material-transfer agreements (MTAs) are used to define the terms under which proprietary material and/or information is exchanged. Neither rights in intellectual property nor rights for commercial purposes are granted under this type of agreement. MTAs typically state that the recipient may use the materials for their own research purposes only and not for any commercial purposes.

Research-Use Exemption Is Practiced as Rational Forbearance

Bristol-Myers Squibb often licenses to the research community basic research tools that we develop that would not otherwise be available to investigators at academic or other research institutions. I am sure that our practice and that of others in the pharmaceutical industry is similar to that of many academic and other research institutions conducting federally funded research. In that context, the National Institutes of Health (NIH) maintains a policy of facilitating the availability of unique or novel biologic materials and resources developed with NIH funds. In all those circumstances, the competing interests must be balanced.

To some extent, intellectual property law helps to provide some rationality for the use of patented research tools. Damages generally cannot be collected from an infringer who is merely engaging in research. Typically, in fact, an inhouse research program is not sufficiently far along to know whether a lawsuit would actually protect valuable property or technology or even whether that property will ultimately prove to be of no value. Frankly, we all know that it is not good form to sue researchers in academic institutions and stifle their progress. Consequently, much potential litigation has been held in check, and we have not often had to confront the vexing issues that would arise in the litigation context. I hope that this rational forbearance will continue.

Why Intellectual Property Is Important in Molecular Biology

Clearly, there are likely to be a variety of important issues on the horizon. In fact, in the slightly more than 30 years that I have been a member of the human genetics community, our field has been responsible for a successive list of the issues that seem to vex the public in ethical, legal, and commercial ways. Early on, it was a prenatal diagnosis, then it was gene therapy, then it was genetic counseling, then it became DNA sequencing, and today it is genomics. Tomorrow, it will surely be something else. That is part of why it has been so exciting to be a member of this community for all these years.

Those issues and others are indicative of the fact that intellectual property rights will be an increasingly important component of future research developments. However, they also demonstrate that neither intellectual property rights nor science can or should try to trump the other. We must continue to engage in reasonable discourse, acknowledge and deal with our differences constructively, and strive to find the compromises that provide maximal support for a biomedical research enterprise that has enormous potential for the alleviation of human suffering. This remains, despite all the problems, a remarkably exciting time for the conduct of biomedical science. Judging by what we heard in this workshop, I have no doubt that it will remain as exciting a time for the commercialization of science as well.

SMALL BIOTECHNOLOGY COMPANY

Tom D'Alonzo, Genvec, Inc.

Genvec's Intellectual Property

What we do at Genvec is start with a virus and declaw it. We do that by pulling out part of its genome and putting human DNA in its place. Then we grow it in commercial quantities that represent possible products. In developing the intellectual property of our business, we concentrate on what to take out of the virus and what to put back into the virus, and perhaps most important—having removed part of its genome—we need to figure out what is required in cell lines to replicate the modified virus, that is, how to grow it. Our objective is to produce a clinical preparation that will meet Food and Drug Administration specifications for use in humans. It is a fairly daunting exercise. The progress that has been made on this in just the last couple of years is striking. I think it is naive to consider gene therapy in any context other than traditional drug development. We should expect a series of generational improvements, in terms of filing patents, in the products that are developed.

Ultimately, we are looking at the clinical application of our virus technology. For us, it has initially been in cystic fibrosis. Within the next few weeks, we hope to start a clinical trial with colon cancer metastatic to the liver through which we expect to improve the performance characteristics of the vectors being used for gene therapy and to add to our understanding of how these vectors work.

Role of Venture Capital

Venture capital is a greater asset in the United States than anywhere else in the world. It provides the engine for the biotechnology companies that are revving up across the United States. In the biotechnology industry, venture capitalists perform the service of identifying potential technologies that might be too underdeveloped, too underadvertised to attract the interest of the larger companies, or too far outside their technology area to induce them to displace ongoing research programs and businesses. A venture capitalist sees that as an opportunity that begins with assessment of the technology. Are there other technologies with which it might compete? How does it relate to the overall market? What about the proprietary estate that surrounds it? Is it something that can be developed, progressed, and finally made into a business? More than any other country in the world, the United States has the ability to pull in venture capital from various sources, tie it up for 10 years with the understanding that there are going to be occasional conspicuous failures, and return enough of a gain to a pension fund or investor to attract more money to invest in the cycle. Venture capitalists take on the risk and invest in a small company. They enlist people to envision how the business is going to be formed and developed. They might spend a bit more money, perhaps raise another round or two of financing, and begin to investigate the commercial application. How is the product going to be made? How is it going to be put into a clinical development program, and what toxicology will support it? Finally, and fairly important for the ability to partner up this technology, can the product be efficiently manufactured in commercial quantities?

Then the discussion with the larger corporate partners begins. For businesses like ours in the biotechnology industry, without the opportunity to establish partnerships with larger companies, we could not raise the money required to bring the technology to the point of availability to a patient as an approved product. The biotechnology companies are conspicuously and unquestionably dependent on larger corporate partners to take products forward through the clinic and into a commercial setting. The 1995 rate of partnering between pharmaceutical companies and biotechnology companies was double the rate in the year before.

At the end of the day, venture capitalists' goal is to realize a profit. Typically, it is through an initial public offering (IPO), which is a marker of the success of a business. The venture capital system is both self-pruning and self-regenerating. It is estimated that there are upwards of 2,000 biotechnology companies in the United States; they represent a market capitalization of more than $60 billion, of which $6-7 billion a year is spent on research. Thus, it is an essential and very successful national experiment that has so far yielded about 30 biotechnology products that have reached the market. To be sure, there are failures along the way. But if we were the kind of nation or if venture capitalists were the kind of people that would not take on a risk for fear of failure, we would not have any of those products. It is more risk and reward that drive this system, rather than a frank fear of failure. The system has provided an incredible benefit to the medical community in the United States.

Issues in Partnering Biotechnology and Pharmaceutical Firms

There are only so many ways that a company at an early stage of development like ours can fund itself. We either raise more money privately, develop our business to the point of the IPO, or find a corporate partner to fund development of the technology. Mergers and acquisitions are another source of funds. Two companies might come together and form a stronger business than either one was before. Alternatively, one company might be acquired by another, in what could be a camouflage for a partial failure (there is no way of knowing without knowledge of specific details about the companies). Mergers or acquisitions themselves can be a source of funding.

From our discussions with several companies, some things have become apparent and predictable about corporate partnering. First is the demand for high-quality science that is going to form the basis for the business. A second is the challenges that still remain with the technology. What is the quality of the people who are behind the challenges? What are the strategies that are being put in place to address them? What is the probability of success?

Finally, a very important consideration is, What do you own, what will we have a right to, and will we have the freedom to operate? If we can look at what you have and envision a product possibility and a pathway to get there, will we have the freedom to commercialize the product under the proprietary estate that surrounds it?

From the industry point of view, some of the trends I see in patents are helpful and some are disturbing. We see broad, sweeping patents at the front end of new technology that begin to narrow in scope as later patent applications are filed. The broad earlier filing can be troublesome, and the later filings become more narrow and specific as the technology matures. Broad earlier filing, if granted, will complicate the commercialization of new technology.

Summary

I take great pride in the fact that industry has delivered a considerable benefit to the health care community. The early venture capitalists who are taking the high risks with the newest, least-developed technology need to be able to see that risk rewarded if they are to sustain the challenge of developing that technology and putting their capital at risk. During its initial phase, a biotechnology company must develop both its business and its technology. When a company finally offers its technology to a corporate partner, the proprietary estate is a prominent piece of what is being offered to the larger corporate partner, and it has an appropriate and a correct expectation in that regard. If there is uncertainty or lack of predictability and confidence in the partnering system, it will complicate the discussions and make forming partnerships much more difficult. We have enough uncertainty on the science side without introducing uncertainty on the patenting side. For the patent system to work well, predictability and consistency must be a part of it.

GOVERNMENT

Harold Varmus, National Institutes of Health

Decisions that the National Institutes of Health (NIH) makes about intellectual property—be it research tools or anything else—are influenced by at least three goals:

- *Fostering scientific discovery.* This includes providing various kinds of incentives to our investigators but making sure that we maintain the health and the integrity of the entire research community that we support with our nearly $12 billion budget.

- *Making sure that the discoveries made by NIH investigators are sufficiently used to foster human health.* This includes our grantees at universities and research institutions, and investigators at the NIH campus. Research information must be transferred to industry so that the public derives the benefit from the tax dollars used to support research to improve strategies for preventing and curing diseases.
- *Protecting the rights of NIH employees when they make discoveries.* Although this goal is sometimes in conflict with the first two goals, as an employer of several thousand scientists at NIH, we must take their rights into consideration when we make decisions about how we protect intellectual property.

Claims have been made in some quarters that, under my direction, NIH is somehow opposed to patenting and licensing. That is clearly untrue. We acknowledge the many benefits of patents and of licenses, such as mandating disclosure, speeding the application of research results to human health, and providing incentives to scientists and discoverers. Since the inception of our Office of Technology Transfer, we have had a very high patenting rate. Nearly 90% of NIH invention disclosures were submitted for patent protection. We are now trying to reduce that rate, both to restrict our costs of patent-claim development and to restrict our applications to inventions that we expect to be most useful. We currently file patent claims on about 60% of our invention reports. That number will probably come down further in the future.

I have been involved in a number of conflicts over intellectual property and research tools. Before I came to NIH, I was deeply involved in the conflict over the sharing of genetically manipulated mice [For further discussion, see National Research Council 1994]. That episode shows how influences of the marketplace and open discussion of issues can lead to solutions that work. The issue was joined initially because one company had attempted to extract fairly large amounts of money from academic investigators for access to mice for which there were patent claims; but most academic investigators felt that they should have access to these mice because their development was sponsored with public funds.

In the course of trying to deal with this somewhat abstract argument, it became apparent that no single company could take in and do the husbandry on a large enough number of strains of genetically altered mice to satisfy the community and that the companies that were trying to do so were not making a profit. Moreover, the scientific community was determined to get an efficient process for distribution of mice. The Jackson Laboratory was sufficiently interested in being the vehicle for distribution, and it handles the mice efficiently. Virtually all genetically altered mice are now available, and most of us do not feel that there is a major problem any longer. There are sometimes some licensing restrictions; but in general, the problem is not acute.

The second episode that has concerned me involved the patent applications that NIH had made before my arrival at NIH on so-called expressed-sequence tags (ESTs). We decided not to appeal the rejection of NIH's application by the Patent and Trademark Office for several reasons: my concern about the lack of demonstrated utility of these sequences; the possible complications of having what is referred to as "patent clutter," that is, multiple patents that would ultimately prove to be held on the same gene; and the problem of speculative claims, or so-called "gotcha" patents, in which someone would do a lot of work on a gene and find that a patent had already been established on the gene. All in all, such patent activity might well restrict progress. Although NIH withdrew from this argument under those circumstances, the issue is not completely resolved. As Bill Haseltine, from Human Genome Sciences, suggested during the workshop, some companies have in fact made patent applications on ESTs. My view is that widespread patenting of ESTs will pose some fairly serious problems because of some of the reasons mentioned above.

A third concern is the agreement that Human Genome Sciences (HGS) was requiring for use of its EST database.[2] The HGS agreement was to allow reasonably free access to what they call level I sequences (sequences that already had appeared in the Genbank database). There was a different kind of requirement for access to so-called level II sequences (those not in Genbank). That requirement was developed after a series of discussions among HGS, NIH, and the Howard Hughes Medical Institute. Scientists using information in the level II database would be required to report discoveries to HGS, maintain strict confidentiality about the sequences, and give HGS options to intellectual property rights considerably downstream of discoveries made by using the database.

Although the HGS agreement is certainly legal, I was not enthusiastic about having either investigators on the NIH campus or academic scientists who are supported by NIH grants become involved in it. I saw the restraints on the abilities of those scientists to communicate freely with their colleagues as unreasonable. I was also concerned about what seemed to be excessive, long-term reach-through provisions. For example, someone who had gotten a sequence from HGS would have to honor the agreement, even if the same sequence had appeared in Genbank two weeks later from an academic source. We did not prohibit our investigators from entering into agreements with HGS, but we did caution them about the restraints that the agreements would impose on them. Our

[2] Discussants at the workshop referred to this as the "HGS agreement." In discussions with NRC staff after the workshop, William Haseltine described this as a "TIGR agreement." The written agreement included two appendixes. One was signed by TIGR (The Institute for Genome Research) and the other by HGS. Because both appendixes extensively reference each, both TIGR and HGS were considered to be signatories to the full agreement.

intramural scientists discussed these issues, and they felt fairly strongly that the secrecy issues and the reach-through provisions were sufficiently troubling for the intramural program to choose not to make use of any agreements with HGS.

We acknowledge that a company like HGS needs to make some profit on its investment in ESTs. Here is one place where self-interest obviously creates a difference of opinion about what should be done. It is interesting to contrast how ESTs are offered to individual investigators with how restriction enzymes are offered. Both are products of companies. If we buy a restriction enzyme for $50 or $250, there might or might not be a patent on it. We are pleased to pay the asking price because it is an efficient way for us to get a useful research tool. No one attempts to seize downstream rights on anything that we do with that restriction enzyme. In contrast, the reach-through provision attached to the use of ESTs creates a very serious problem for us, and that (tied in with the secrecy arrangements) has discouraged us from pursuing the EST databank that is made available through level II agreements.

What should be patented, and what should be placed in the public domain? I see this as the central question. Approaches to such questions are influenced by three major factors: self-interest, the investment of the public in government-supported research, and the law. How do we determine whether our answers to the questions are correct? With respect to what is legal, one must simply wait for decisions from either the judiciary or the Patent and Trademark Office. The more interesting issue is how we judge which decisions to make with respect to the goals of NIH and the interests of the public. It is sometimes easier for a company to judge whether it has made the right decision. Did it make a profit? Did it get more investment capital? A university, in some cases, might try to weigh the effectiveness of what it does on the basis of whether there are royalty returns. But, as Lita Nelsen pointed out, there are many other reasons why a university would want to get involved in technology transfer and intellectual property protection—reasons that are much harder to measure.

From NIH's perspective, how do we measure whether the scientific community is more productive as a result of decisions that we have made? How do we measure whether applications of knowledge occur more efficiently? These are difficult questions that we would like to know the answers to, but we do not.

One way to look at this is to follow the outcome of the use of sequences that are in the HGS-TIGR database (TIGR, The Institute for Genome Research is a nonprofit partner of HGS), as opposed to sequences that have been put into the public database as a result of the activities of Merck. Will sequences that investigators obtain only by going through level II agreements with HGS produce more benefits than those studied by academic investigators and obtained free of any attachments from Genbank after being sequenced by Washington University and paid for by Merck? This would be a useful experiment.

One of the things that we learn from these discussions is that every example of a research tool that breeds contention has its own characteristics and its own solutions. Nevertheless, in looking at the history of how our science is developed and how we use the available instruments of intellectual property protection, it would be of interest for many of us to know, even in retrospect, whether placing restraints on information and attempting to exploit our ability to restrict information for the benefit of one party or another actually has any public benefit.

REFERENCES

Blumenthal D, Causino N, Campbell E, and Louis KS. 1996. Relationships between academic institutions and industry in the life sciences—an industry survey. New England J Med. 334(6):368-373.

National Research Council. 1994. Sharing Laboratory Resources: Genetically Altered Mice. Washington. National Academy Press. 41p.

7

Summary

The workshop discussions covered much ground and many beliefs about the advantages and disadvantages of different strategies for protecting intellectual property. Many common assumptions about the economic benefits of patenting do not withstand close scrutiny, and none applies across the board to research tools in molecular biology. Although in many peoples' minds protection of intellectual property raises a question of whether to patent, controversies over licensing agreements and protection of unpatented intellectual property seem to be more common. The consensus among workshop participants was that the most important issue is *access* to, not patenting of, research tools. Participants from the three sectors—academe, industry, and government—made it clear that they are not opposed to patenting of research tools. By and large, they expressed the view that broad access to research tools is important for the continued vigor of the research enterprise and that research in molecular biology is in the public interest because of its potential to contribute to human health. At the same time, the specter of patents that could stifle research permeated many of the discussions. Even though they have not been patented en masse, the patenting of espressed sequence tags (ESTs) seems to represent the worst fears of many participants. Indeed, the issue was raised in almost every case study and repeatedly in general discussion. Many participants raised the concern that patents on genes or gene fragments could result in fragmented ownership that would severely disrupt scientific progress, or as Bennett Shapiro put it, "restricting access this way could Balkanize the human genome."

Discussions among the workshop participants made it clear that one should not expect to identify a single guiding principle for the protection of intellectual property surrounding research tools. The differences in circumstances surrounding the various case studies argue against easy generalizations. Even the recognition that a particular innovation is an important research tool does not define clear strategies. Not only does the category have no legal status, it is not even a discrete category. A technology might be a research tool in the hands of some users and a means of production in the hands of others. Even as a research tool, the technology takes different forms. Participants drew a distinction between research aids and discovery tools. That is, a research tool can be more like a screwdriver—designed to turn screws of a specific size and shape—or more like a telescope—a discovery tool useful for scanning horizons. Describing something as a research tool generally describes only particular instances of its use and almost never defines its full range of uses or values.

Leon Rosenberg noted that the research-use exemption is used by many companies and described it as a "rational forbearance," which he hoped would continue. Many workshop participants, especially those working in universities, expressed surprise that the research-use exemption was not an established legal principle. However, until this exemption is upheld in a court of law, it will remain simply a general practice with no legal protection and will be sustained primarily by individuals' self-interest.

All the speakers at the workshop associated with major pharmaceutical companies—Bennet Shapiro, Thomas Caskey, and Leon Rosenberg—mentioned that it was in their organizations' self-interest to disseminate their research tools broadly. In contrast, speakers from smaller biotechnology firms noted that their survival depended on their ability to disseminate research tools on a more restricted basis. Both sectors claim a common goal—to develop useful applications from advances in molecular biology—but the means by which they can achieve this goal, given their business constraints, are quite different.

Finally, the variation in the technology of different research tools argues against the wisdom of searching for universal guiding principles. Perhaps the only guiding principle that one might derive from categorizing a particular innovation as a research tool is the observation that the innovation has the potential to foster scientific progress. Certainly, none of the workshop participants questioned the principle that scientific progress in biomedical research should not be impeded, no matter how the intellectual property rights are distributed. Instead, the disputes about rights of access to research tools lay in the details. Does the delay in disclosure that results from patent application inhibit scientific progress? Is access to research tools broad enough to facilitate scientific progress? Do the benefits of a protected environment for further development outweigh the costs of excluding others from this research tool?

Indeed, rather than providing answers, the most practical contribution of the workshop might have been to suggest questions. What sorts of issues might technology transfer managers consider in deciding how to manage research tools? The following questions emerged from our distillation of the common themes raised by workshop participants and from comparisons of the case studies. They might be useful for informing decisions about appropriate strategies for protecting intellectual property that constitutes important research tools.

- Is this technology critical to researchers? Are there other ways to do the same thing?
- Is private investment necessary to make the research tool more useful? Is the size of the potential market sufficient to warrant the amount of extra investment?
- Does or might this research technology have alternative uses for which commercial markets exist, and would private investment be necessary to develop those markets?
- Are multiple dissemination strategies feasible (such as different licensing terms and different fees)?
- Where substantial further investment is necessary to develop the technology, will further development yield additional patentable inventions sufficient to motivate investment?
- Is this invention likely to be only one of many pieces to a complex puzzle, each of which could be separately owned?

Those questions encapsulate themes raised by many of the speakers. Many participants spoke of the problem of uncertainty, which is likely to be a continuing problem in setting policies about proprietary decisions. To some degree, it seems that uncertainty is an inevitable byproduct of the recent advances in molecular biology. Molecular biology has already presented the patent system with new categories of patentable material, for which case law—which of necessity looks to the past for guidance—cannot always be expected to provide adequate guidance. One can expect advances in molecular biology to continue to pose new and unexpected challenges to our intellectual property traditions. As science continues to present new challenges, case law will inevitably lag behind the science. Because the useful applications of advances in molecular biology are so closely tied to basic research, it is particularly important to strike the right balance between the twin goals of fostering the growth of scientific knowledge and promoting private investment in the development of applications for new technology. In many cases, those goals can be encouraged by the same strategies; in others, the strategies might conflict. The balance is not likely to be achieved through any single policy and clearly requires the close and continued attention of people concerned with the science and application of molecular biology.

Agenda of the February 1996 Workhop

OPENING REMARKS	Eric Fischer, *Board on Biology, National Research Council*
Patenting Research Tools and the Law	Rebecca Eisenberg, *University of Michigan*
What Should be Public and What Should be Private in Science and Technology?	Richard Nelson, *Columbia University*
Sharing Intellectual Property: Empirical Studies of Scientific Exchange Among Human Genome Researchers	Steven Hilgartner, *Cornell University*

CASE STUDY DISCUSSIONS

Recombinant DNA	*Moderator:* Sidney Winter, *Wharton School of Business* Floyd Grolle, *Stanford University* Lita Nelsen, *Massachusetts Inst. Technology*

Protein and DNA Sequencing	*Moderator:* Robert Cook-Deegan, *National Research Council* Leroy Hood, *University of Washington* Harry Osterman, *LI-CORE, Inc.* Waclaw Szybalski, *University of Wisconsin*
Gene Fragments (ESTs)	*Moderator*: Gerald Rubin, *UC-Berkeley* William Haseltine, *Human Genome Sciences* Thomas Caskey, *Merck Research Laboratories* James Sikela, *University of Colorado*
PCR and Taq Polymerase	*Moderator:* Barbara Mazur, *DuPont Nemours Co.* Ellen Daniell, *Roche Molecular Systems* Bernie Poiesz, *University of Syracuse* Tom Gallegos, *Oncorpharm*
Research Tools in Drug Discovery	*Moderator:* William Comer, *SIBIA Inc.* Bennett Shapiro, *Merck Research Laboratories* Larry Respess, *Ligand Pharmaceuticals, Inc.* James Wilson, *Inst. for Human Genome Therapy*
Case Study Summaries: Overview of Issues Raised in Case Studies	Janet Joy*, *Board on Biology, National Research Council*

PERSPECTIVES FROM DIFFERENT SECTORS

What practices have worked well? What have not ?	*Moderator:* Ronald Sederoff, *North Carolina State University*

* Gene Block, Vice Provost for Research and Professor of Biology at the University of Virginia, was originally scheduled to give this presentation but was unable to attend.

University	Gerald Rubin, *University of California at Berkeley* Lita Nelsen, *Massachusetts Institute of Technology*
Private Sector	Leon Rosenberg, *Bristol-Myers Squibb* Thomas D'Alonzo, *Genvec, Inc.*
Government	Harold Varmus, *National Institutes of Health*
General Discussion	*Moderator:* Ronald Sederoff